SpringerBriefs in Applied Sciences and Technology

Computational Intelligence

Series editor

Janusz Kacprzyk, Warsaw, Poland

About this Series

The series "Studies in Computational Intelligence" (SCI) publishes new developments and advances in the various areas of computational intelligence—quickly and with a high quality. The intent is to cover the theory, applications, and design methods of computational intelligence, as embedded in the fields of engineering, computer science, physics and life sciences, as well as the methodologies behind them. The series contains monographs, lecture notes and edited volumes in computational intelligence spanning the areas of neural networks, connectionist systems, genetic algorithms, evolutionary computation, artificial intelligence, cellular automata, self-organizing systems, soft computing, fuzzy systems, and hybrid intelligent systems. Of particular value to both the contributors and the readership are the short publication timeframe and the world-wide distribution, which enable both wide and rapid dissemination of research output.

More information about this series at http://www.springer.com/series/10618

Daniela Sanchez · Patricia Melin

Hierarchical Modular Granular Neural Networks with Fuzzy Aggregation

 Springer

Daniela Sanchez
Division of Graduate Studies
Tijuana Institute of Technology
Tijuana, Baja California
Mexico

Patricia Melin
Division of Graduate Studies
Tijuana Institute of Technology
Tijuana, Baja California
Mexico

ISSN 2191-530X   ISSN 2191-5318   (electronic)
SpringerBriefs in Applied Sciences and Technology
ISBN 978-3-319-28861-1   ISBN 978-3-319-28862-8   (eBook)
DOI 10.1007/978-3-319-28862-8

Library of Congress Control Number: 2015960430

Printed on acid-free paper

This Springer imprint is published by SpringerNature
The registered company is Springer International Publishing AG Switzerland

Preface

In this book, a new model of modular neural network based on a granular approach, the combination of their responses, and the optimization by hierarchical genetic algorithms are introduced. The new model of modular neural networks is applied to human recognition, and for this four databases of biometric measures are used; face, iris, ear, and voice. The different responses are combined using type-1 and interval type-2 fuzzy logic. Finally, two hierarchical genetic algorithms are used to perform the optimization of the granular modular neural networks parameters and fuzzy inference system parameters. The experimental results obtained using the proposed method show that when the optimization is used, the results can be better than without optimization.

This book is intended to be a reference for scientists and engineers interested in applying soft computing techniques, such as neural networks, fuzzy logic, and genetic algorithms; all of them apply to human recognition, but also in general to pattern recognition and hybrid intelligent systems and similar ones. We consider that this book can also be used to find novel ideas for new lines of research, or to continue the lines of research proposed by authors of the book.

In Chap. 1, a brief introduction to the book is presented, where the intelligence techniques that are used, the main contribution, motivations, application, and a general description of the proposed methods are mentioned.

We present in Chap. 2 the background and theory required to understand the methods and ideas presented later in the book. In this chapter some concepts such as modular neural network, fuzzy logic, genetic algorithm, granular computing, and a brief description of the application used for testing our proposed method are explained. This chapter allows readers to understand better the different techniques used in the proposed method.

In Chap. 3, we explain the proposed method, where the new model is detailed by explaining each step needed to achieve our proposed method. In addition, the parameters, equations, and figures are presented to describe deeply our main idea. First, the two proposed granulation methods are described with their optimizations, and second, the combination of responses are described with its respective optimization using hierarchical genetic algorithms.

In Chap. 4, we present a detailed explanation of how the proposed method is applied to human recognition, and the different databases that are used in the experiments are described in a detailed fashion.

We show the experimental results achieved in this research work in Chap. 5, where the different performed tests are described and the best results are shown for each database. The non-optimized tests show the corresponding modular granular neural networks architectures; meanwhile the optimized results show the best results and the graphical convergences of the best optimizations.

In Chap. 6, we offer our general conclusions and future works of this research work. In a general way, a new model of modular granular neural networks is introduced, where the combination of responses is performed using fuzzy logic, and its optimization is also developed to improve the results.

We want to express our heartfelt thanks to our colleague and friend, Prof. Janusz Kacprzyk, for his kindness, support, and motivation to perform and write our research work. We thank the two supporting agencies in our country for their help to develop this research work: CONACYT and DGEST. Also, we thank our institution, Tijuana Institute of Technology, for always supporting our projects. We thank our families, for their support and motivation during the different stages of this research work. Finally, we thank all our colleagues working in soft computing, and all people who have helped, contributed, and motivated us to perform and write this research work, are too many to mention all of them by name.

Tijuana, Mexico   Daniela Sanchez
September 2015   Patricia Melin

Contents

Chapter 1
Introduction

Hybrid intelligent systems are computational systems that integrate different intelligent techniques. This integration is mainly performed because each technique can have some individual limitations, but these limitations can be overcome if these techniques are combined. These systems are now being used to support complex problem solving. These kind of systems allow the representation and manipulation of different types and forms of data and knowledge which may come from various sources. The hybrid intelligent systems are integrated with different techniques, and these techniques are divided into two categories: traditional hard computing techniques and soft computing techniques (such as fuzzy logic, neural networks, and genetic algorithms) [1].

In this book, a new method of hybrid intelligent system is proposed. The proposed method is based on granular computing applied in two levels. The techniques used and combined in the proposed method are modular neural networks (MNNs) with a Granular Computing (GrC) approach, thus resulting in a new concept of MNNs; modular granular neural networks (MGNNs), also fuzzy logic (FL) and hierarchical genetic algorithms (HGAs) are techniques used in this research work. These techniques are chosen because in other works have demonstrated to be a good option, and in the case of MNNs and HGAs, these techniques allow to improve the results obtained than with their conventional versions; respectively artificial neural networks and genetic algorithm [2–4].

The main contribution in this research work is to develop a new method to perform the granulation in 2 levels. In the first level, a granulation of the data is performed. Smaller sub granules are obtained of this granulation and each sub module learns a sub granule, i.e. a part of a main granule. The proposed method is applied to human recognition and the granulation of number of data for training is performed, this granulation is the same for all the sub granules but the granulation of the number of persons in each sub granule can be different. This model can be developed "N" times, i.e., for "N" number of biometric measure can be used and each biometric measure has a different granulation.

© The Author(s) 2016
D. Sanchez and P. Melin, *Hierarchical Modular Granular Neural Networks with Fuzzy Aggregation*, SpringerBriefs in Computational Intelligence, DOI 10.1007/978-3-319-28862-8_1

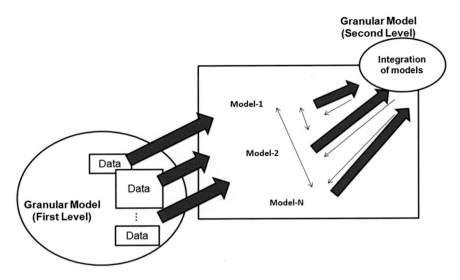

Fig. 1.1 Proposed granulation

In the second level of granulation, the combination of MGNNs responses is performed, and in this case, fuzzy logic is used. The number of responses to be integrated depends on the number of biometric measures used. In this part, the number of membership functions have an important role among other parameters used, that have an important impact in the final result. Optimizations in both levels are performed to improve the results obtained. In Fig. 1.1, an illustration of the granulation is shown.

The development of the proposed method is mainly motivated, because in previous works, the kind of proposed granulation has not been used. Usually, when a modular or ensemble neural network is used, the number of sub modules and data for training are fixed. If we talk about human recognition, the number of persons in each module is also fixed and even, the number of persons is equitably divided between the sub modules. When the responses of the modular or ensemble neural networks are integrated using a fuzzy inference system, the number of the membership functions in the fuzzy integrators is also fixed in each variable [4–6]. The proposed granulation has as main goal the change in the division of data, parameters and other information than other works establish in a fixed way.

This book is organized as follows: In Chap. 2, the background and theory are presented. The description of the proposed method is presented in Chap. 3. In Chap. 4, the application to prove the effectiveness of the proposed method and the databases used are described. In Chap. 5, the experimental results are shown. The conclusions are presented in Chap. 6.

References

1. Zhang, Z., Zhang, C.: An agent-based hybrid intelligent system for financial investment planning. PRICAI 355–364 (2002)
2. Hidalgo, D., Melin, P., Licea, G., Castillo, O.: Optimization of type-2 fuzzy integration in modular neural networks using an evolutionary method with applications in multimodal biometry. MICAI 454–465 (2009)
3. Melin, P., Felix, C., Castillo, O.: Face recognition using modular neural networks and the fuzzy Sugeno integral for response integration. Int. J. Intell. Syst. **20**(2), 29–275 (2005)
4. Muñoz, R., Castillo, O., Melin, P.: Face, fingerprint and voice recognition with modular neural networks and fuzzy integration. Bio-inspired Hybrid Intell. Syst. Image Anal. Pattern Recogn. 69–79 (2009)
5. Hidalgo, D., Castillo, O., Melin, P.: Optimization with genetic algorithms of modular neural networks using interval type-2 fuzzy logic for response integration: the case of multimodal biometry. IJCNN 738–745 (2008)
6. Melin, P., Sánchez, D., Castillo, O.: Genetic optimization of modular neural networks with fuzzy response integration for human recognition. Inf. Sci. **197**, 1–19 (2012)

Chapter 2
Background and Theory

In this chapter a brief overview of the basic concepts used in this research work is presented.

2.1 Human Recognition

Establishing the identity of an individual is very important in our society. The primordial task of any identity management system is the ability to determine or validate the identity of its users prior to granting them access to the resources protected by the system. Usually, passwords and ID cards have been used to validate the identity of an individual intending to access the services offered by an application or area. However, such mechanisms for user authentication have several limitations. For example, the passwords can be divulged to unauthorized users and the ID cards can be forged or stolen. An alternative is the use of biometric systems. Biometrics is the science of establishing the identity of an individual based on the inherent physical or behavioral traits associated with the person [1]. Biometric systems utilize fingerprints, iris, face, hand geometry, palm print, ear, voice, signature, among others, in order to recognize exactly a person [2].

2.2 Artificial Neural Network

Neural networks (NNs) can be used to extract patterns and detect trends that are too complex to be noticed by either humans or other computer techniques [3, 4]. The term of non-optimized trainings is used when the architecture of a conventional neural network, an ensemble neural network or a modular neural network, is established randomly or by trial and error, i.e. a method to find an optimal architecture is not used. Some parameters of this kind of neural networks can be the

© The Author(s) 2016
D. Sanchez and P. Melin, *Hierarchical Modular Granular Neural Networks
with Fuzzy Aggregation*, SpringerBriefs in Computational Intelligence,
DOI 10.1007/978-3-319-28862-8_2

number of hidden layers, the number of neurons for each hidden layer, the learning algorithm, number of epoch, among others [5–7]. This kind of NNS has two phases; the training phase (learning phase) and the testing phase. Depending of the information given in the training phase, the output performance is affected [8].

2.2.1 Modular Neural Networks

The concept of modularity is an extension of the principle of divide and conquer, the problem should be divided into smaller sub problems that are solved by experts (sub modules) and their partial solutions should be integrated to produce a final solution [9]. The results of the different applications involving Modular Neural Networks (MNNs) lead to the general evidence that the use of modular neural networks implies a significant learning improvement comparatively to a single NN and especially to the back-propagation NNs. Each neural network works independently in its own domain. Each of the neural networks is built and trained for a specific task [10, 11]. In this research work, modular neural networks are used, where each database or dataset is learned by a neural network of this kind, taking the advantage provided by the modularity.

2.3 Granular Computing

The term of Granular computing was introduced in 1997 [12, 13] and has been studied in different areas such as; computational intelligence, artificial intelligence, data mining, social networks among others [14–16]. For example in [17], the privacy protection in social network data based on granular computing is considered. In that work, the granulation consists of grouping individuals with the same combination of attribute values into a common pool or information granule [18].

The main difference between granular computing and data clustering techniques is that the second ones offer a way of finding relationships between datasets and grouping data into classes. A basic task of granular computing is to build a hierarchical model for a complex system or problem [13, 19]. The basic ingredients of granular computing are granules, a web of granules, and granular structures. A granule can be interpreted as one of the numerous small particles forming a larger unit [19–24]. At least three basic properties of granules are needed: internal properties reflecting the interaction of elements inside a granule, external properties revealing its interaction with other granules and, contextual properties showing the relative existence of a granule in a particular environment. A complex problem is represented as a web of granules, in which a granule may itself be a web of smaller granules. The granule structures represent the internal structures of a compound granule. The structures are determined by its element granules in terms of their composition, organization, and relationships. Each element granule is simply considered as a single point when studying the structure of a compound granule. From the viewpoint of element granules, the internal structures of a compound granule are indeed their collective structures. Each granule in the family captures and represents a particular and local aspect of the problem [25–29]. In this research work, a granulation approach is used in 2 levels. In the first level, a granulation of a dataset or database is performed and the information is learned using modular neural networks and, in the second level also this approach is used when the responses of the modular neural networks are combined.

2.4   Fuzzy Logic

Fuzzy logic is an area of soft computing that enables a computer system to reason with uncertainty [30]. A fuzzy inference system consists of a set of if-then rules defined over fuzzy sets. Fuzzy sets generalize the concept of a traditional set by allowing the membership degree to be any value between 0 and 1 [31]. This corresponds, in the real world to many situations, where it is difficult to decide in an unambiguous manner if something belongs or not to a specific class [11]. Fuzzy logic is a useful tool for modeling complex systems and deriving useful fuzzy relations or rules [32]. However, it is often difficult for human experts to define the fuzzy sets and fuzzy rules used by these systems [33]. The basic structure of a fuzzy inference system consists of three conceptual components: a rule base, which contains a selection of fuzzy rules, a database (or dictionary) which defines the membership functions used in the rules, and a reasoning mechanism that performs the inference procedure [34].

2.4.1   Type-2 Fuzzy Logic

The concept of a type-2 fuzzy set was introduced as an extension of the concept of an ordinary fuzzy set (henceforth now called a "type-1 fuzzy set"). A type-2 fuzzy set is characterized by a fuzzy membership function, i.e., the membership grade for each element of this set is a fuzzy set in [0,1], unlike a type-1 set where the membership grade is a crisp number in [0,1]. Such sets can be used in situations where there is uncertainty about the membership grades themselves, e.g., an uncertainty in the shape of the membership function or in some of its parameters. Consider the transition from ordinary sets to fuzzy sets. When we cannot determine the membership of an element in a set as 0 or 1, we use fuzzy sets of type-1. Similarly, when the situation is so fuzzy that we have trouble determining the membership grade even as a crisp number in [0,1], we use fuzzy sets of type-2 [8, 35–37]. Uncertainty in the primary memberships of a type-2 fuzzy set, $\tilde{A}$, consists of a bounded region that we call the "footprint of uncertainty" (FOU). Mathematically, it is the union of all primary membership functions [38–41].

In this book, type-1 and type-2 fuzzy logic are used, because both types of fuzzy logic are a good option to perform the combination of responses when modular neural networks are used and, because the type of fuzzy logic used must depend basing on the complexity of the problem to be solved for this reason both types of fuzzy logic are considered. The optimization of type of fuzzy logic and their parameters are optimized to focus on the needs of each case or problem.

2.5 Genetic Algorithms

A Genetic Algorithm (GA) is an optimization and search technique based on the principles of genetics and natural selection. A GA allows a population composed of many individuals to evolve under specified selection rules to a state that maximizes the "fitness" [42, 43]. Genetic Algorithms (GAs) are nondeterministic methods that use crossover and mutation operators for deriving offspring. GAs work by maintaining a constant-sized population of candidate solutions known as individuals (chromosomes) [44–46]. Competent genetic algorithms can efficiently address problems in which the order of the linkage is limited to some small number k, called order-k separable problems. The class of problems with high-order linkage cannot be addressed efficiently in general. However, specific subclasses of this class can be addressed efficiently if there is some form of structure that can be exploited. A prominent case is given by the class of hierarchical problems [47].

2.5.1 Hierarchical Genetic Algorithms

Introduced in [48], a Hierarchical genetic algorithm (HGA) is a particular type of genetic algorithm. Its structure is more flexible than the conventional GA. The basic idea under a hierarchical genetic algorithm is that for some complex systems, which cannot be easily represented, this type of GA can be a better choice. The complicated chromosomes may provide a good new way to solve the problem and have demonstrated to achieve better results in complex problems than the conventional genetic algorithms [49, 50]. The hierarchical genetic algorithms have genes dominated by other genes, this means than depending of the value of the major genes or control genes will be the behavior of the rest of the genes [51]. This kind of genetic algorithm is used in this research work because, it allows the activation or deactivation of genes (parametric genes) depending on the behavior of the control genes. This advantage is exploited in the proposed method in the optimizations performed, where the number of sub modules or sub granules (when the optimization of modular neural network architectures is performed) and the type of fuzzy logic (when the optimization of fuzzy inference systems parameters is performed) are control genes.

References

1. Jain, A.K., Bolle, R.M., Pankanti, S.: Biometrics: Personal Identification in Networked Society. Springer; Edition: 1st ed. 1999. 2nd printing, 2005
2. Ross, A., Jain, A.K.: Human recognition using biometrics: an overview. Annales des Telecommunications **62**(1–2), 11–35 (2007)
3. Azamm, F.: Biologically Inspired Modular Neural Networks. PhD thesis, Virginia Polytechnic Institute and State University, Blacksburg, Virginia (2000)
4. Khan, A., Bandopadhyaya, T., Sharma, S.: Classification of stocks using self organizing map. Int. J. Soft Comput. Appl. **4**, 19–24 (2009)
5. Melin, P., Sánchez, D., Castillo, O.: Genetic optimization of modular neural networks with fuzzy response integration for human recognition. Inf. Sci. **197**, 1–19 (2012)
6. Melin, P., Urias, J., Solano, D., Soto, M., Lopez, M., Castillo, O.: Voice recognition with neural networks, type-2 fuzzy logic and genetic algorithms. J. Eng. Lett. **13**(2), 108–116 (2006)
7. Valdez, F., Melin, P., Parra, H.: Parallel genetic algorithms for optimization of Modular Neural Networks in pattern recognition. IJCNN 314–319 (2011)
8. Hidalgo, D., Melin, P., Licea, G., Castillo, O.: Optimization of type-2 fuzzy integration in modular neural networks using an evolutionary method with applications in multimodal biometry. MICAI 454–465 (2009)
9. Santos, J.M., Alexandre, L.A., Marques de Sá, J.: Modular neural network task decomposition via entropic clustering. ISDA **1**, 62–67 (2006)
10. Auda, G., Kamel, M.S.: Modular neural networks a survey. Int. J. Neural Syst. **9**(2), 129–151 (1999)
11. Melin, P., Castillo, O.: Hybrid Intelligent Systems for Pattern Recognition Using Soft Computing: An Evolutionary Approach for Neural Networks and Fuzzy Systems, 1st edn., pp. 119–122. Springer (2005)
12. Lin, T.Y.: Granular Computing, Announcement of the BISC Special Interest Group on Granular Computing (1997)
13. Lin, T.Y., Yao, Y.Y., Zadeh, L.A.: Data Mining, Rough Sets and Granular Computing. Physica-Verlag, Heidelberg (2002)
14. Jankowski, A., Skowron, A.: Toward perception based computing: a rough-granular perspective. WImBI **2006**, 122–142 (2006)
15. Pedrycz, W., Skowron, A., Kreinovich, V.: Handbook of Granular Computing Hardcover, 1st edn. Wiley-Interscience (2008)
16. Skowron, A., Stepaniuk, J., Swiniarski, R.W.: Modeling rough granular computing based on approximation spaces. Inf. Sci. **184**(1), 20–43 (2012)
17. Wang, D.W., Liau, C.J., Hsu, T.: Privacy Protection in Social Network Data Disclosure Based on Granular Computing. In: Proceedings of 2006 IEEE International Conference on Fuzzy Systems, IEEE Computer Society, pp. 997–1003 (2006)
18. Nettleton, D.F., Torra, V.: Data privacy for simply anonymized social network logs represented as graphs: considerations for graph alteration operations. Int. J. Uncertain. Fuzziness Knowl. Based Syst. **19**(1), 107–125 (2011)
19. An, A., Stefanowski, J., Ramanna, S., Butz, C.J., Pedrycz, W., Wang, G.: Rough Sets, Fuzzy Sets, Data Mining and Granular Computing. In: 11th International Conference, RSFDGrC 2007, Proceedings. Lecture Notes in Computer Science 4482. Springer (2007)

20. Peters, J.F., Pawlak, Z., Skowron, A.: A rough set approach to measuring information granules. In: Proceedings of COMPSAC, pp. 1135–1139 (2002)
21. Polkowski, L., Artiemjew, P.: A study in granular computing: on classifiers induced from granular reflections of data. Trans. Rough Sets **9**, 230–263 (2008)
22. Polkowski, L., Artiemjew, P.: Granular computing in the frame of rough mereology. A case study: classification of data into decision categories by means of granular reflections of data. Int. J. Intell. Syst. **26**(6), 555–571 (2011)
23. Yao, J.T., Vasilakos, A.V., Pedrycz, W.: Granular computing: perspectives and challenges. IEEE Trans. Cybern. **43**(6), 1977–1989 (2003)
24. Yao, Y.Y.: Granular Computing: Past, Present and Future (GrC), pp. 80–85 (2008)
25. Bargiela, A., Pedrycz, W.: The Roots of Granular Computing. In: IEEE International Conference on Granular Computing (GrC), pp. 806–809 (2006)
26. Yao, Y.Y.: On Modeling Data Mining with Granular Computing. In: 25th International Computer Software and Applications Conference, (COMPSAC), pp. 638–649 (2001)
27. Yao, Y.Y.: Perspectives of Granular Computing. In: IEEE International Conference on Granular Computing (GrC), pp. 85–90 (2005)
28. Yu, F., Pedrycz, W.: The design of fuzzy information granules: tradeoffs between specificity and experimental evidence. Appl. Soft Comput. **9**(1), 27–264 (2009)
29. Zadeh, L.A.: Some reflections on soft computing, granular computing and their roles in the conception, design and utilization of information/intelligent systems. Soft. Comput. **2**, 23–25 (1998)
30. Castillo, O., Melin, P.: Soft Computing for Control of Non-Linear Dynamical Systems. Springer, Heidelberg (2001)
31. Zadeh, L.A.: Fuzzy sets. J. Inf. Control **8**, 338–353 (1965)
32. Okamura, M., Kikuchi, H., Yager, R., Nakanishi, S.: Character diagnosis of fuzzy systems by genetic algorithm and fuzzy inference. In: Proceedings of the Vietnam-Japan Bilateral Symposium on Fuzzy Systems and Applications, Halong Bay, Vietnam, pp. 468–473 (1998)
33. Wang, W., Bridges, S.: Genetic Algorithm Optimization of Membership Functions for Mining Fuzzy Association Rules. Department of Computer Science Mississippi State University (2000)
34. Jang, J., Sun, C., Mizutani, E.: Neuro-Fuzzy and Soft Computing. Prentice Hall, New Jersey (1997)
35. Hidalgo, D., Castillo, O., Melin, P.: Optimization with genetic algorithms of modular neural networks using interval type-2 fuzzy logic for response integration: the case of multimodal biometry. IJCNN 738–745 (2008)
36. Hidalgo, D., Castillo, O., Melin, P.: Type-1 and type-2 fuzzy inference systems as integration methods in modular neural networks for multimodal biometry and its optimization with genetic algorithms. Soft Comput. Hybrid Intell. Syst. 89–114 (2008)
37. Sánchez, D., Melin, P.: Modular neural network with fuzzy integration and its optimization using genetic algorithms for human recognition based on iris, ear and voice biometrics. In: Soft Computing for Recognition Based on Biometrics Studies in Computational Intelligence, 1st edn., pp. 85–102. Springer, (2010)
38. Castillo, O., Melin, P.: Type-2 Fuzzy Logic Theory and Applications, pp. 29–43. Springer, Berlin (2008)
39. Castro, J.R., Castillo, O., Melin, P.: An interval type-2 fuzzy logic toolbox for control applications. FUZZ-IEEE 1–6 (2007)
40. Castro, J.R., Castillo, O., Melin, P., Rodriguez-Diaz, A.: Building fuzzy inference systems with a new interval type-2 fuzzy logic toolbox. Trans. Comput. Sci. **1**, 104–114 (2008)
41. Mendel, J.: Uncertain Rule-Based Fuzzy Logic Systems: Introduction and New Directions. Upper Saddle River, Prentice-Hall (2001)
42. Haupt, R., Haupt, S.: Practical Genetic Algorithms, 2nd edn., pp. 42–43. Wiley-Interscience (2004)
43. Mitchell, M.: An Introduction to Genetic Algorithms. A Bradford Book, 3rd edn (1998)

44. Coley, A.: An Introduction to Genetic Algorithms for Scientists and Engineers. Wspc, Har/Dskt edition (1999)
45. Huang, J., Wechsler, H.: Eye Location Using Genetic Algorithm. In: Second International Conference on Audio and Video-Based Biometric Person Authentication, pp. 130–135 (1999)
46. Nawa, N., Takeshi, F., Hashiyama, T., Uchikawa, Y.: A study on the discovery of relevant fuzzy rules using pseudobacterial genetic algorithm. IEEE Trans. Ind. Electron. **46**(6), 1080–1089 (1999)
47. Jong, E., Thierens, D., Watson, R.: Hierarchical Genetic Algorithms. In: The 8th International Conference on Parallel Problem Solving from Nature (PPSN), pp. 232–241 (2004)
48. Tang, K.S., Man, K.F., Kwong, S., Liu, Z.F.: Minimal fuzzy memberships and rule using hierarchical genetic algorithms. IEEE Trans. Ind. Electron. **45**(1), 162–169 (1998)
49. Wang, C., Soh, Y.C., Wang, H., Wang, H.: A hierarchical genetic algorithm for path planning in a static environment with obstacles. In: Canadian Conference on Electrical and Computer Engineering, 2002. IEEE CCECE 2002, vol. 3, pp. 1652–1657 (2002)
50. Worapradya, K., Pratishthananda, S.: Fuzzy supervisory PI controller using hierarchical genetic algorithms. In: 5th Asian Control Conference, 2004, vol. 3, pp. 1523–1528 (2004)
51. Sánchez, D., Melin, P.: Optimization of modular granular neural networks using hierarchical genetic algorithms for human recognition using the ear biometric measure. Eng. Appl. Artif. Intell. **27**, 41–56 (2014)

Chapter 3
Proposed Method

In this chapter, the proposed method is presented. The proposed method combines modular granular neural networks and fuzzy logic integration. The main idea of the proposed method is to have "**N**" Sub datasets or Databases and each of these is divided into modular granular neural networks and each modular granular neural network can have "**_m_**" sub granules or sub modules. The different responses of each MGNN are combined using a fuzzy integrator. The number of inputs in the fuzzy integrator depends on the number of modular granular neural networks used. To improve the performance of the proposed method, two hierarchical algorithms are proposed. The first HGA is for the optimization of the modular granular neural network architectures and the second for the optimization of parameters of the fuzzy integrators. In Fig. 3.1, the general architecture of the proposed method is presented.

3.1 Granulations

In this section, the description of the different proposed granulations in this research work and their optimizations are presented. In the first granulation, 3 backpropagation algorithms are used to perform the modular granular neural networks simulations because in others works the best results are achieved with these algorithms [1–4]. The 3 fastest algorithms considered in this research work are:

1. Gradient descent with scaled conjugate gradient (SCG): This algorithm updates weight and bias values according to the scaled conjugate gradient method. This algorithm may require more iterations to converge than other algorithms of this kind, but its advantage is that in each iteration reduces significantly the number of computations.
2. Gradient descent with adaptive learning and momentum (GDX): this algorithm updates weight and bias values according to gradient descent momentum and an adaptive learning rate.

© The Author(s) 2016
D. Sanchez and P. Melin, *Hierarchical Modular Granular Neural Networks with Fuzzy Aggregation*, SpringerBriefs in Computational Intelligence,
DOI 10.1007/978-3-319-28862-8_3

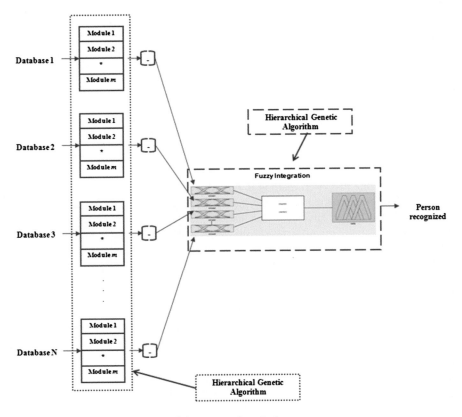

Fig. 3.1 The general architecture of the proposed method

3. Gradient descent with adaptive learning (GDA): The algorithm performance is very sensitive to the learning rate value. The performance can be improved if the learning rate can change during the training process.

Other parameters for the modular granular neural network simulations must be considered. The minimum and maximum values used for the parameters are shown in Table 3.1, but if more modules or hidden layers are needed, only these parameters must be changed, due to the fact that proposed method is developed to change its architecture depending on how many modules or hidden layers are needed.

Table 3.1 Table of values

Parameters of MNNs	Minimum	Maximum
Modules (*m*)	1	10
Percentage of data for training	20	80
Error goal	0.000001	0.001
Learning algorithm	1	3
Hidden layers (*h*)	1	5
Neurons for each hidden layers	20	400
Epoch	–	2000

3.1.1 First Granulation

The main idea is to find how many modules are needed depending on the database. The main granule is the whole database, and this database is divided into different number of sub granules (from 1 to *m*), each sub granules can have different size, i.e. different number of data. In the case of human recognition, each sub granule will have different number of persons. In Fig. 3.2, the proposed granulation method is illustrated.

In the case of human recognition, a method to select which data or samples are used for the training and testing phases based on the percentage of data for training is also proposed. Depending on the number of data for training (***nod***) equivalents to the percentage of data for training, the method randomly selects which images or sample will be used for each phase. An example is shown in Fig. 3.3, where the total number of data (***tnod***) is equal to 5, and randomly the data #1, #4 and #5 are selected to the training phase and data #2 and #3 for testing.

There are other techniques for the selection of the data or samples for the training phase, some examples of these are used in bagging (bootstrap) [5] and boosting techniques. These techniques are used to improve the accuracy of classification and their selection of data also consists into obtain randomly data for the training phase [6].

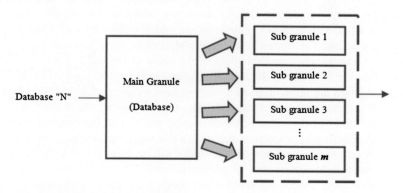

Fig. 3.2 The proposed granulation method

Percentage of data for training: 60%
Number of data for training (*nod*): 3
Training: 1 , 4 and 5
Testing: 2 and 3

Fig. 3.3 Example of training and test data selection

The main difference between these techniques and the proposed method for data selection is that the number of data or samples used in the training phase (***nod***) in bagging and boosting is equal to the total number of data (***tnod***), i.e. some data or samples are used more than once (are repeated) for the training phase and the main idea of the proposed method for selection of data is to reduce the data used in the training phase.

3.1.1.1 Optimizations for the First Granulation

The optimization of the first granulation is performed to find optimal modular granular neural network architectures, i.e. which is the optimization of the granulation previously described using a hierarchical genetic algorithm. As mentioned previously, the main idea of the proposed method is to know how many sub granules or modules are needed to obtain good results and, for this reason a hierarchical genetic algorithm is proposed in the method. The main advantage of the hierarchical genetic algorithm (HGA) above the conventional genetic algorithms (GA) is to work with complex optimizations, which cannot be easily represented with a simple GA. The hierarchical genetic algorithms have genes dominated by other genes, this means than depending of the value of the major genes or control genes will be the behavior of the rest of the genes (active or inactive). In this HGA, the control genes are being used to determine how many modules or sub granules are needed and how much percentage of data for the training phase is needed, and depending mainly of the value of the first gene, the rest of the genes (parametric genes) will be activated or inactive, this means that the genes will be used or will be simply ignored, and so, different architectures will be created and with the fitness function, those architectures will be evaluated.

This hierarchical genetic algorithm allows finding an optimal architecture, which means that the hierarchical genetic algorithm performs the optimization of various parameters of the architecture of a modular granular neural network, such as the number of sub granules or modules, percentage of data for training, error goal, learning algorithm, number of hidden layers and their respective number of neurons. The proposed hierarchical genetic algorithm used the conventional elitism. This means that in each generation the best individual of the population (in this case with the best architecture) is saved and separated of the rest of the population to avoid that this individual be modified when the genetic operators are used.

The chromosome of this hierarchical genetic algorithm is shown in Fig. 3.4. This algorithm seeks to minimize the recognition error and the fitness function is given by Equation:

$$f1 = \sum_{i=1}^{m} \left(\left(\sum_{j=1}^{n_m} X_j \right) \Big/ n_m \right) \qquad (3.1)$$

where m is the total number of modules, X_j is 0 if the module provides the correct result and 1 if not, and n_m is total number of data points used for testing in the corresponding module.

The size of the chromosome can be calculated as:

$$Size = 2 + (3 * m) + (m * h) \qquad (3.2)$$

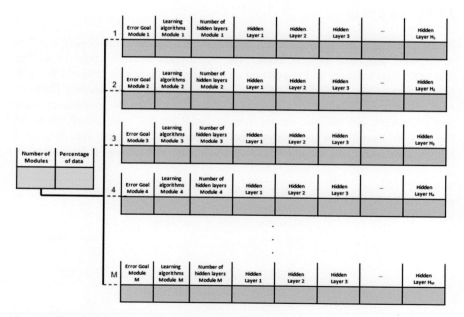

Fig. 3.4 Chromosome of the proposed HGA for MNNs

Where **m** is the total number of sub granules or modules that the hierarchical genetic algorithm can use, and **h** is total number of hidden layers per module that the hierarchical genetic algorithm can use. The size of the chromosome is of 132 genes. For this application, two stopping conditions are established: when the maximum number of generations is achieved and when the best individual has an error value equal to zero.

These variables, **m** and **h** can be established depending of the application or the database. The proposed method is tested using the parameters shown in Table 3.2, and some of these parameters are based on the parameters recommended in [7]. For mutation, the breeder genetic algorithm (BGA) is used, this algorithm uses the selected individuals (the number of individuals depends on the selection rate). These individuals are randomly mutated (based on the mutation rate) and the resulting population is returned and recombined with the rest of the population [7].

3.1.2   Second Granulation (Based on Database Complexity)

The second granulation is proposed to be used with big databases using a granular approach. This method has 2 stages; first, the use of nonoptimized trainings, and second, based on the results achieved in the first stage, a grouping is performed to obtain the optimal design of the MGNNs.

3.1.2.1   First Stage

In the first stage, the non-optimized trainings must be performed using the first granulation proposed in this book. The main idea of this stage is to use non-optimized trainings and to perform an analysis of the results obtained for each person, that means, depending on how many data are used for the training phase, in each of the non-optimized trainings, we need to know when the modular granular neural network did not provide the correct result (in this case called error). In Fig. 3.5, an example of the analysis can be observed, when five non-optimized trainings are used, and the number of errors for the first person is determined in

Table 3.2 Table of parameters for the HGA for the MGNNs optimization

Genetic operator	Value
Population size	10
Maximum number of generations	30
Selection	Roulette wheel
Selection rate	0.85
Crossover	Single point
Crossover rate	0.9
Mutation	BGA
Mutation rate	0.01

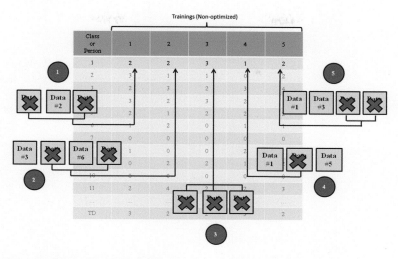

Fig. 3.5 Example of the analysis

each of the five trainings. This process must be performed for all the persons from $P_1, P_2, ..., P_d$ *where d* is the number of persons.

How to perform this analysis is described below. First, the amount of data for the training phase in each training is stored, an example of this is presented in Table 3.3, where five trainings are established, but the number of trainings can be established according to any situation.

The total of data for training (*tdt*) is given by Equation:

$$tdt = \sum_{t=1}^{tr} dt_t \qquad (3.3)$$

where *tr* is the total of training non-optimized used, dt_t is the number of data for testing used in the training *t*.

A matrix is created, because each data for training of each training is evaluated with their respective modular granular neural network (for each person), if the modular granular neural network provides the correct result is a 0 and 1 if not. An example of this matrix can be observed in Table 3.4 where the value of *d* is equal to 12.

The percentage of complexity (*poc*) is given by Equation:

$$poc = 100 * \left(\sum_{i=1}^{tdt} x_i \right) / tdt \qquad (3.4)$$

Table 3.3 Example of number of data for training

Training (*t*)	1	2	3	4	5
Data for testing (*dt*)	3	4	3	3	4

Table 3.4 Table of zeros and ones

Person	Trainings (non-optimized)														
	1			2			3			4			5		
1	1	1	0	1	0	1	1	1	0	1	0	0	0	1	1
2	1	1	0	0	1	0	0	0	0	0	0	1	1	0	0
3	1	0	1	1	1	0	1	0	1	1	1	1	1	1	1
4	1	1	0	1	0	1	1	1	1	0	1	0	1	0	0
5	0	1	1	0	0	0	1	1	1	0	1	0	1	1	0
6	0	0	1	0	1	0	0	0	1	0	0	0	1	0	0
7	0	0	0	0	0	0	0	0	0	0	0	0	0	0	0
8	1	0	0	0	0	0	0	0	1	1	0	0	0	0	1
9	0	0	0	1	1	0	1	0	0	1	0	1	1	0	0
10	0	0	0	0	0	0	0	0	0	0	0	0	0	0	0
11	1	1	1	1	1	1	1	1	0	1	1	1	1	0	1
12	1	1	0	0	1	1	1	0	1	1	1	1	0	1	0

where **tdt** is the total of data for training of all the trainings used, X_i is 0 if the modular granular neural network provided the correct result and 1 if not. This equation is used for all the persons (P_1, P_2 ,..., P_d), and an example of the calculation of the percentage of complexity (**poc**) is presented in the last column in Table 3.5.

A summary of results for the errors (of each training) and the percentage of complexity (**poc**) is shown in Table 3.6

When the percentage of complexity (**poc**) of each person is calculated, the rules are established. These rules are freely created based on the percentage of complexity (**poc**) and a complexity level using labels is established depending of that percentage, because the main idea is to perform a granulation based on the error obtained in the non-optimized trainings, an example of these rules can be observed in Fig. 3.6.

When the rules are established, each person is evaluated using its percentage of complexity and a complexity level is assigned. An example of this can be observed in the last column of Table 3.7.

When all the persons have a complexity level, the persons are ordered in descending order based on its percentage of complexity (**poc**), and depending of its position in that order a new number of identification (Number of ID) is assigned to each person. In Table 3.8, an example is shown.

Then, automatically depending on the number of rules that were established then the number of sub granules are created, grouping the persons with the same complexity level. In Table 3.9, an example is presented.

The persons with the same complexity level are shown on Table 3.10.

In Fig. 3.7, a summary of the process performed up to this moment is presented:

(1) A database is used.
(2) Non-optimized trainings are performed (with random parameters), and based on rules the complexity levels are assigned to each person.
(3) A new number of ID is assigned to each person, depending on its complexity level, and the persons with the same complexity level are grouped. The number of sub granules will depend on the number of rules.
(4) How many modules will be needed to each sub granule?. This question is answered with a hierarchical genetic algorithm described below.

3.1.2.2 Second Stage (Optimization)

For the second stage, a hierarchical genetic algorithm is proposed, and one of the objectives of this HGA is to find out how many modules are needed in each sub granule. In Fig. 3.8, a summary of all the processes can be observed. The last process is to perform the optimization.

This HGA is described in this section and shown in Fig. 3.9.

The optimization is the last process for the second granulation. The main idea of the optimization is to find out the number of sub granules or modules that are

Table 3.5 Example of calculation of *poc*

Person	Trainings (non-optimized)																	Percentage of complexity (*poc*)
	1			2			3				4				5			
1	0	1	0	1	1	0	1	1	1	1	0	1	0	0	0	0	1	58.82 %
2	1	1	1	0	0	1	0	0	1	0	0	0	0	1	0	1	0	41.17 %
3	1	1	1	1	1	0	0	1	1	0	1	1	1	1	1	1	1	82.35 %
4	1	1	1	0	1	1	1	1	1	1	1	0	1	1	1	0	0	64.70 %
5	1	1	1	1	0	0	0	1	1	1	1	0	1	1	1	1	0	52.94 %
6	1	0	0	0	1	0	0	0	0	0	1	0	0	1	1	0	0	29.41 %
7	0	0	0	0	0	0	0	0	0	0	0	0	0	0	0	0	0	0
8	0	1	1	0	0	0	0	0	0	0	1	1	0	0	0	0	1	23.52 %
9	0	0	0	0	1	0	0	1	1	0	0	1	0	1	0	0	0	41.17 %
10	0	0	0	0	0	0	0	0	0	0	0	0	0	0	0	0	0	0
11	0	1	0	1	1	1	1	1	1	0	0	1	1	1	1	0	1	76.47 %
12	1	1	1	1	0	1	1	0	1	1	1	1	1	0	0	1	0	70.58 %

Table 3.6 Summary of results (example of calculation of poc)

Person	Trainings (non-optimized)					Percentage of complexity (*poc*)
	1	2	3	4	5	
1	2	2	3	1	2	58.82 %
2	3	1	1	0	2	41.17 %
3	2	3	2	3	4	82.35 %
4	3	2	3	2	1	64.70 %
5	2	1	2	2	2	52.94 %
6	1	2	0	1	1	29.41 %
7	0	0	0	0	0	0
8	1	0	0	2	1	23.52 %
9	0	2	2	1	2	41.17 %
10	0	0	0	0	0	0
11	2	4	2	2	3	76.47 %
12	3	2	2	3	2	70.58 %

needed in each sub granule of the process #3 (See Fig. 3.8). As method to perform the optimization, a hierarchical genetic algorithm (HGA) was chosen to allow the activation or deactivation of genes (parametric genes) depending on the behavior of the control genes. This advantage is exploited in the proposed method because when certain condition is fulfilled and a control gene is going to be disabled, this last one performs a deactivation of genes that depend on that gene and the chromosome size is changed.

This hierarchical genetic algorithm performs the optimization of the modular granular neural networks architectures, its work consists in finding some optimal parameters as the first genetic algorithm previously presented. These parameters are the number of modules of each sub granule, error goal, learning algorithm, number of hidden layers and number of neurons. Mainly, the first gene will determine the

Fig. 3.6 Example of the rules

If *poc* >= 65 then
                        Complexity Level = High
Else

If *poc* >= 35 and *poc* <= 65 then
                        Complexity Level = Medium
Else

If *poc* > 0 and *poc* <= 35 then
                        Complexity Level = Low
Else

If *poc* = 0 then
                        Complexity Level = Zero
End

Table 3.7 Example of the complexity level

Person	Error	*tdt*	Percentage of complexity (*poc*)	Complexity level
1	10	17	58.82 %	Medium
2	7	17	41.17 %	Medium
3	14	17	82.35 %	High
4	11	17	64.70 %	Medium
5	9	17	52.94 %	Medium
6	5	17	29.41 %	Low
7	0	17	0	Zero
8	4	17	23.52 %	Low
9	7	17	41.17 %	Medium
10	0	17	0	Zero
11	13	17	76.47 %	High
12	12	17	70.58 %	High

number of validation to optimize. This HGA has 4 methods of elitism and deactivation processes, a re-initialization process and stopping conditions, and all these are described later. In Fig. 3.10, the chromosome of this hierarchical genetic algorithm is shown. The objective of this algorithm is to minimize the recognition error and the fitness function is given by Eq. 3.1, the same used for the first HGA of this book.

The size of the chromosome can be calculated as:

$$Size = 1 + ((m * 3) + (m * h) * \mathbf{g}) \tag{3.5}$$

where m is the total number of modules that can be used by the HGA, and h is total number of hidden layers per module that can be use by the HGA and $\mathbf{g}$ is the

Table 3.8 Example of new number of ID

Person	Error	*tdt*	Percentage of complexity (*poc*)	Complexity level	New number of ID
1	10	17	58.82 %	Medium	5
2	7	17	41.17 %	Medium	8
3	14	17	82.35 %	High	1
4	11	17	64.70 %	Medium	4
5	9	17	52.94 %	Medium	6
6	5	17	29.41 %	Low	9
7	0	17	0	Zero	11
8	4	17	23.52 %	Low	10
9	7	17	41.17 %	Medium	7
10	0	17	0	Zero	12
11	13	17	76.47 %	High	2
12	12	17	70.58 %	High	3

Table 3.9 Example of reordering

New number of ID	Percentage of complexity (*poc*)	Complexity level
1	82.35 %	High
2	76.47 %	High
3	70.58 %	High
4	64.70 %	Medium
5	58.82 %	Medium
6	52.94 %	Medium
7	41.17 %	Medium
8	41.17 %	Medium
9	29.41 %	Low
10	23.52 %	Low
11	0	Zero
12	0	Zero

number of granules established in process #3. The *m* and *h* values are shown in Table 3.1, and **g** will depend on the number of rules. The genetic operators values used for the experimental results are shown in Table 3.2.

Elitisms

The methods used for elitism are described below. The first elitism method is always used in each generation, the rest of the elitism methods are randomly selected in each generation.

First Elitism

The first elitism is an external memory and the main idea of this external memory is to save the best result (that part of the chromosome, the modular granular neural network, and the error value) of each granule and each validation, averages (by validation and granule) and a general average is calculated. First, the number of possible validations will be calculated by Equation:

$$V_{MAX} = \sum_{r=1}^{tdm} \frac{n!}{r!(n-r)!} \tag{3.6}$$

Table 3.10 Granulation non-optimized (example)

Granule	Number of person	Complexity level
1	3	High
2	5	Medium
3	2	Low
4	2	Zero

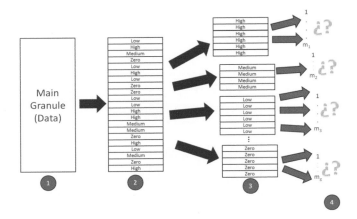

Fig. 3.7 Process of the first stage

where **n** is the total of data points of the database (images per person), **r** is the number of data for training, and **tdm** is the maximum number of data for the training phase. For example if **n** is equal to 5, and **tdm** is equal to 4, then V_{max} is equal to 30, and for this example, the possible validations are presented in Table 3.11.

An example of this external memory is shown in Fig. 3.11, where each validation has a space in the external memory, but for big databases it would seem impossible to evaluate in only one evolution all the possible validations. For this reason when each evolution starts, it randomly selects which validations will be used in that evolution.

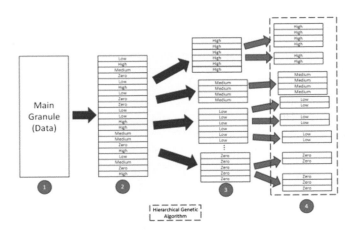

Fig. 3.8 Summary of all the processes

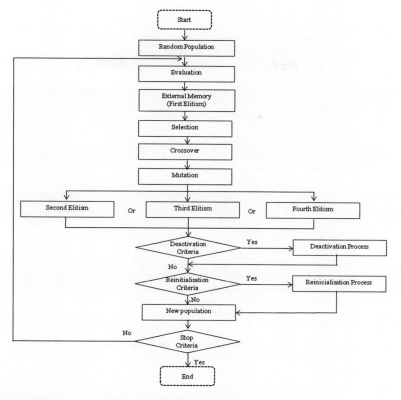

Fig. 3.9 Diagram of the HGA for the second granulation

The number of validations used in the evolution is easily established, logically the number of validation used must be less or equal to V_{max}, when this number is established, and which validation will be used is selected randomly (all the validation will be different), in Fig. 3.12, a real example of external memory of an evolution can be observed using 5 validations during the evolution.

The same external memory and its evolution are shown in Fig. 3.13. A difference between Figs. 3.12 and 3.13 can be clearly observed. The main advantage of this method of elitism versus the conventional is that for example if during the evolution an individual of the HGA has a bad objective function value, this individual will be usually eliminated, and using the proposed external memory if that individual has good results in one or more granule (compared with the stored in the external memory), that part of the chromosome is saved, and thus allows the evolution of each validation and each granule.

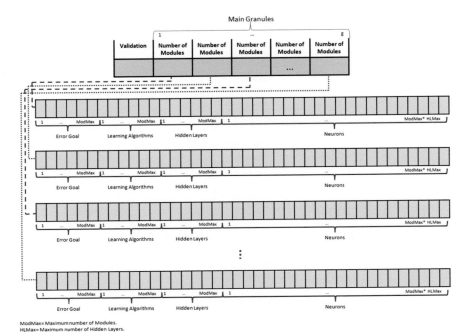

Fig. 3.10 Chromosome of the proposed HGA for the second granulation

Second Elitism

The second method of elitism used in the proposed method is the conventional method, where the individual with the best performance is saved to avoid being modified with the genetic operators. An example of this method of elitism is shown in Fig. 3.14, where an example of the population can be observed (in this case the objective function is of minimization).

Third Elitism

The third method of elitism also uses the population, but in this case the best result (and that part of the chromosome) of each granule is saved and a new individual is generated, but the validation with the worst value of objective function is assigned to that individual. An example of this method of elitism is shown in Fig. 3.15.

Fourth Elitism

The external memory is used by the fourth method of elitism, the best result (and that part of the chromosome) of each granule is saved and a new individual is generated, but the validation with the worst value of the objective function is assigned to that individual. An example of this method of elitism is shown in Fig. 3.16.

Table 3.11 Table of validations

	Data for Training				
	Data for Testing				
V1	1	2	3	4	5
V2	2	1	3	4	5
V3	3	1	2	4	5
V4	4	1	2	3	5
V5	5	1	2	3	4
V6	1	2	3	4	5
V7	1	3	2	4	5
V8	1	4	2	3	5
V9	1	5	2	3	4
V10	2	3	1	4	5
V11	2	4	1	3	5
V12	2	5	1	3	4
V13	3	4	1	2	5
V14	3	5	1	2	4
V15	4	5	1	2	3
V16	1	2	3	4	5
V17	1	2	4	3	5
V18	1	2	5	3	4
V19	1	3	4	2	5
V20	1	3	5	2	4
V21	1	4	5	2	3
V22	2	3	4	1	5
V23	2	3	5	1	4
V24	2	4	5	1	3
V25	3	4	5	1	2
V26	1	2	3	4	5
V27	1	2	3	5	4
V28	1	2	4	5	3
V29	1	3	4	5	2
V30	2	3	4	5	1

Validation / Granule	V_1	V_2	V_3	...	V_{MAX}	Average
1						
2						
3						
⋮						
Granules						
Average						

Fig. 3.11 Example of external memory

Validation Granule	V_{s1}	V_{s2}	V_{s3}	V_{s4}	V_{s5}	Average
1	0.1747	0.2795	0.0577	0.0551	0.1058	**0.1346**
2	0.2271	0.1711	0.0968	0.0283	0.1486	**0.1344**
3	0.0143	0.0175	0.0027	0.0016	0	**0.0072**
Average	**0.1387**	**0.1560**	**0.0524**	**0.0284**	**0.0848**	**0.0921**

Fig. 3.12 Real Example of external memory (Part #1)

Deactivation of Granules or Validations and Stopping Conditions

In Fig. 3.17, the deactivation processes are identified in an example of the external memory. The main idea of these processes is that when each granule with all the validations used (1) and each validation with all the granules (2) can have an average error as goal (In Fig. 3.13, the average of error as goal is equal to zero, and this error was only achieved by the V_{s5}), and when this average error is obtained, this granule or validation stops its evolution and is deactivated, and the evolution will be concentrated in the rest of granules or validations. Another deactivation process occurs when some granule of any validation (3) has an error equal to zero, that granule is deactivated in that validation. When one or more granules are deactivated, but the rest of the granules are still working, depending on the validation that will be evaluated, the modular neural networks that are represented in this validation in the deactivated granules are taken of the external memory and then only with the activated granules the trainings are performed, and the simulation is performed with all de granules with the respective validation. The proposed hierarchical genetic algorithm has two stopping conditions. The first one can occur when an established general average is obtained (4), and the second when the number of generations is achieved.

Validation Granule	V_{s1}	V_{s2}	V_{s3}	V_{s4}	V_{s5}	Average
1	0.1877	0.0761	0.0712	0.0283	0.1063	**0.0939**
2	0.1306	0.0761	0.0459	0.0551	0.1058	**0.0827**
3	0.001	0	0	0	0	**0.0002**
Average	**0.1065**	**0.0507**	**0.039**	**0.0278**	**0.0707**	**0.0589**

Fig. 3.13 Real example of external memory (Part #2)

Individual	V_s	Granule 1	Granule 2	Granule 3	Error
1	3	0.0577	0.0968	0.0027	**0.0524**
2	1	0.1747	0.2271	0.0143	0.1387
3	2	0.9709	0.1711	0.0459	0.3960
4	5	0.1058	0.1486	0.0041	0.0862
5	2	0.2795	0.2186	0.0175	0.1719
6	4	1	0.9961	1	0.9987
7	3	0.8632	0.9721	0.9454	0.9269
8	5	0.2019	0.187	0	0.1296
9	1	0.4407	0.4208	0.0338	0.2984
10	3	0.2511	0.2851	0.0301	0.1887

Best Individual | 3 | 0.0577 | 0.0968 | 0.0027 |

Fig. 3.14 Example of the second elitism method

In this case the goal error for the deactivation processes is equal to zero. As a summary, the deactivation processes occur when:

(1) An average of error of recognition in any granule using all its validations is achieved.
(2) An average of error of recognition in any validation using all its granules is achieved.
(3) An average of error of recognition in any granule of any validation is achieved.
(4) A general average is achieved.

Individual	V_s	Granule 1	Granule 2	Granule 3	Error
1	3	**0.0577**	**0.0968**	0.0027	0.0524
2	1	0.1747	0.2271	0.0143	0.1387
3	2	0.9709	0.1711	0.0459	0.3960
4	5	0.1058	0.1486	0.0041	0.0862
5	2	0.2795	0.2186	0.0175	0.1719
6	**4**	1	0.9961	1	0.9987
7	3	0.8632	0.9721	0.9454	0.9269
8	5	0.2019	0.187	**0**	0.1296
9	1	0.4407	0.4208	0.0338	0.2984
10	3	0.2511	0.2851	0.0301	0.1887

New Individual | 4 | 0.0577 | 0.0968 | 0 |

Fig. 3.15 Example of the third elitism method

Validation Granule	V_{s1}	V_{s2}	V_{s3}	V_{s4}	V_{s5}	Average
1	0.1747	0.2795	0.0577	0.0551	0.1058	0.1346
2	0.2271	0.1711	0.0968	0.0283	0.1486	0.1344
3	0.0143	0.0175	0.0027	0.0016	0	0.0072
Average	0.1387	0.1560	0.0524	0.0284	0.0848	0.0921

| New Individual | 2 | 0.0551 | 0.0283 | | 0 | |

Fig. 3.16 Example of the fourth elitism method

Re-initialization Process

All the selected validations must be evaluated approximately the same number of times, if not; all the population in the first gene (the validation gene) will be again randomly generated. It is important to say, that only the genes will generated with the validations that are under the average of evaluations. In Fig. 3.18, an example of the number of evaluations are shown, where with the criterion, the validation genes will be only generated with validation V_{s4} and V_{s5}. This process is activated only in odd generations.

3.2 Fuzzy Integration

The proposed method in this research work combines the responses of modular granular neural networks (MGNNs) previously presented using the fuzzy logic as response integrators. Each modular granular neural network represents an input in the fuzzy integrator. The idea is the use of "N" number of inputs (depending on the

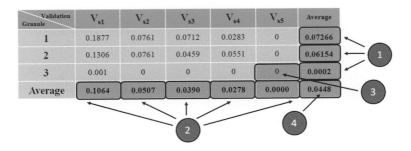

Validation Granule	V_{s1}	V_{s2}	V_{s3}	V_{s4}	V_{s5}	Average
1	0.1877	0.0761	0.0712	0.0283	0	0.07266
2	0.1306	0.0761	0.0459	0.0551	0	0.06154
3	0.001	0	0	0	0	0.0002
Average	0.1064	0.0507	0.0390	0.0278	0.0000	0.0448

Fig. 3.17 Examples of deactivation

	V_{s1}	V_{s2}	V_{s3}	V_{s4}	V_{s5}	Average of Evaluations
Evaluations	7	11	8	3	1	6

Fig. 3.18 Example of re-initialization process

responses to be combined, i.e., the number of modular granular neural networks, sub granules or sub modules) and a final output will be obtained, in this case represented with an output in the fuzzy integrator. An example of the fuzzy inference system is shown in Fig. 3.19.

3.2.1 Optimization of Fuzzy Inference Systems

The fuzzy inference systems optimization is performed because to find good results is difficult if we do not know the correct parameters. In this book another hierarchical genetic algorithm is proposed, but in this case to perform the optimization of some parameters of the fuzzy inference systems. The proposed HGA performs the optimization of the type of fuzzy logic (type-1, interval type-2 fuzzy logic), type of system (Mamdani or Sugeno), type of membership functions (Trapezoidal or gBell), number of membership functions in each variable (inputs and output), their parameters, the consequents of the fuzzy rules and the fuzzy rules. The chromosome of the proposed hierarchical genetic algorithm for the fuzzy inference systems is shown in Fig. 3.20.

3.2.1.1 Optimization of Membership Functions

The number of membership functions in each variable (inputs and output) are optimized, also the type of these membership functions (MFs), but usually when the optimization of the type of membership functions is performed all the membership functions of all the variables are of the same type. In the proposed HGA for the

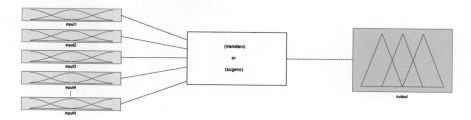

Fig. 3.19 Example of the fuzzy inference system

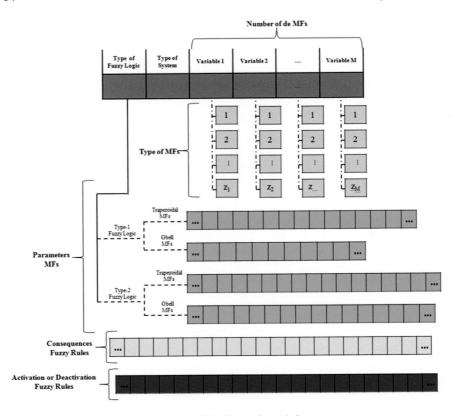

Fig. 3.20 Chromosome of the proposed HGA for the fuzzy inference systems

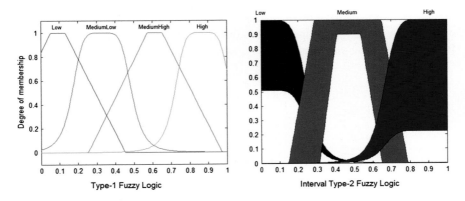

Fig. 3.21 Example of the fuzzy inference system optimization

Table 3.12 Minimum and maximum values of MFs

Parameter	Value
Minimum number of MFs in each variable (MINMFs)	2
Maximum number of MFs in each variable (MNMFs)	5

fuzzy inference system optimization this part is different, because each MFs of each variable (inputs and output) is optimized, i.e. the combination of MFs in the same variable can be possible depending on the type of fuzzy logic. In this Book, only 2 types of MFs are used (Trapezoidal and GBell). In Fig. 3.21, an example can be found.

This optimization performs the fuzzy rules design in two parts; the consequents and the number of fuzzy rules. To perform the consequents optimization, depending of the maximum number of membership function (MNMFs) used in each variable (this number is freely established before the evolution) all the possible rules are generated. When an individual of the HGA is going to be evaluated, depending of the number of membership functions indicated by the genes for the inputs, the possible rules for this combination are taken with their respective consequents. The consequents are taken depending of the number of membership function indicated by the gen for the output. The number of fuzzy rules is performed using genes that indicate the activation or deactivation of the fuzzy rules (each fuzzy rule has a gen), i.e., if the value of the gene is 0 the fuzzy rule is not used, and if the value is 1 is used, this is for all the possible fuzzy rules depending on the combination indicated by the genes for the inputs. Finally, the resulting fuzzy rules are added to the fuzzy integrator. In this research work, the values for the minimum and maximum number of membership functions are shown in Table 3.12.

The genetic parameters used to test the proposed hierarchical genetic algorithm are shown in Table 3.13.

Table 3.13 Table of parameters for the HGA for the FIS optimization

Genetic operator/parameters	Value
Population size	10
Maximum number of generations	100
Selection	Roulette wheel
Selection rate	0.85
Crossover	Single point
Crossover rate	0.9
Mutation	BGA
Mutation rate	0.01

References

1. Gaxiola, F., Melin, P., Valdez, F., Castillo, O.: Neural network with Type-2 fuzzy weights adjustment for pattern recognition of the human Iris biometrics. MICAI (2) 259–270 (2012)
2. Melin, P., Sánchez, D., Castillo, O.: Genetic optimization of modular neural networks with fuzzy response integration for human recognition. Inf. Sci. **197**, 1–19 (2012)
3. Muñoz, R., Castillo, O., Melin, P.: Face, fingerprint and voice recognition with modular neural networks and fuzzy integration. Bio-inspired Hybrid Intell. Syst. Image Anal. Pattern Recogn. 69–79 (2009)
4. Vázquez, J.C., Lopez, M., Melin, P.: Real time face identification using a neural network approach. Soft Comput. Recogn. Based Biometrics 155–169 (2010)
5. Breiman, L.: Bagging predictors. Mach. Learn. **24**, 123–140 (1996)
6. Sutton, C.D.: Classification and regression trees, bagging, and boosting. Handbook of Statistics **24**, 311–329 (2005)
7. Man, K.F., Tang, K.S., Kwong, S.: Genetic Algorithms: Concepts and Designs. Springer (1999)

Chapter 4
Application to Human Recognition

The proposed method combines modular granular neural networks and fuzzy logic integration and, in this section the applications used to prove the effectiveness of the proposed method are described below.

The main idea of the proposed method applied to human recognition is to have "N" Databases and each of these is divided into modular granular neural networks and each modular neural network can have "m" sub granules or sub modules. The different responses of each modular granular neural network are combined using a fuzzy integrator. The number of inputs in the fuzzy integrator depends on the number of modular granular neural network (biometric measures) used. In Fig. 4.1, the general architecture of the proposed method using the first granulation previously mentioned, applied to human recognition is presented.

To improve the performance of the proposed method, two hierarchical genetic algorithms are proposed. The first HGA is for the optimization of the modular granular neural network architectures, the same HGA is used with less percentage of data for the training phase. Finally the optimization of parameters of the fuzzy integrators is also performed.

4.1 Databases

The different databases used in this research work are described. The first four databases are used applying the first granulation and the fuzzy integration, 77 persons are used of each database. The last database (Essex Database) is used for the second granulation based on the complexity.

© The Author(s) 2016
D. Sanchez and P. Melin, *Hierarchical Modular Granular Neural Networks with Fuzzy Aggregation*, SpringerBriefs in Computational Intelligence, DOI 10.1007/978-3-319-28862-8_4

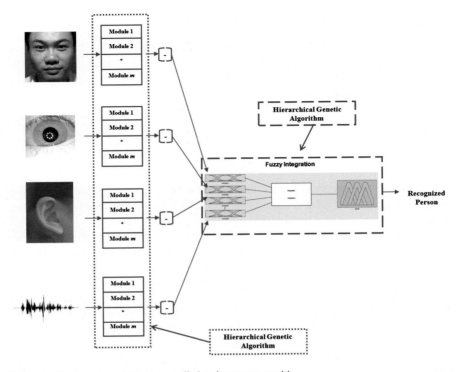

Fig. 4.1 The general architecture applied to human recognition

4.1.1   Face Database (CASIA)

The database of human iris from the Institute of Automation of the Chinese Academy of Sciences is used [1]. Each person has 5 images. The image dimensions are 640 × 480, BMP format. Figure 4.2 shows examples of the human iris images.

Fig. 4.2 Examples of the face images from CASIA database

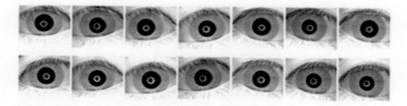

Fig. 4.3 Examples of the human iris images from CASIA database

4.1.2 Iris Database

The database of human iris from the Institute of Automation of the Chinese Academy of Sciences is used [2]. Each person has 14 images (7 for each eye). The image dimensions are 320 × 280, JPEG format. Figure 4.3 shows examples of the human iris images.

4.1.3 Ear Database

The database of the University of Science and Technology of Beijing is used [3]. Each person has 4 images, the image dimensions are 300 × 400 pixels, the format is BMP. Figure 4.4 shows examples of the human ear images.

4.1.4 Voice Database

In the case of voice, the database was collected from students of Tijuana Institute of Technology. Each person has 10 voice samples, WAV format. The word that they said in Spanish was "ACCESAR". To preprocess the voice the Mel Frequency Cepstral Coefficients were used.

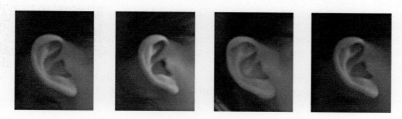

Fig. 4.4 Examples of Ear Recognition Laboratory from USTB

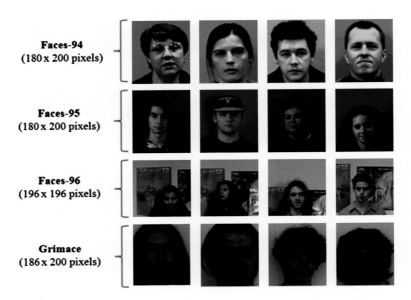

Faces-94
(180 × 200 pixels)

Faces-95
(180 × 200 pixels)

Faces-96
(196 × 196 pixels)

Grimace
(186 × 200 pixels)

Fig. 4.5 Sample of the database from the University of Essex

4.1.5 Face Database (ESSEX)

The database from Essex [4] contains 395 persons divided into 4 directories;
Faces-94 (153 persons), Faces-95 (72 persons), Faces-96 (152 persons) and
Grimace (18 persons), each person has 20 images. The images of a person are
shown in Fig. 4.5.

References

1. Database of Face: Institute of Automation of Chinese Academy of Sciences (CASIA). Found on
   the Web page: http://biometrics.idealtest.org/dbDetailForUser.do?id=9. Accessed 11 Nov 2012
2. Database of Human Iris: Institute of Automation of Chinese Academy of Sciences (CASIA).
   Found on the Web page: http://www.cbsr.ia.ac.cn/english/IrisDatabase.asp. Accessed 21 Sep
   2009
3. Database Ear Recognition Laboratory from the University of Science and Technology Beijing
   (USTB). Found on the Web page: http://www.ustb.edu.cn/resb/en/index.htm.asp. Accessed 21
   Sep 2009
4. Spacek, L.: "Face Recognition Data, University of Essex", UK. 2010, Database ESSEX. Found
   on the Web page: http://cswww.essex.ac.uk/mv/allfaces/. Accessed 21 Nov 2013

Chapter 5
Experimental Results

In this chapter, the best biometric measures results obtained using the first granulation, their integration, the second granulation and the optimizations are presented, the averages can be consulted at the end of this section. The remaining results can be reviewed in the sections of Appendices.

5.1 Modular Granular Neural Networks Results

The results of each biometric measure using the different proposed granulations are presented below.

5.1.1 Granulation #1

The results obtained using the first proposed granulation are presented below, where also a method to perform the optimization of the modular granular neural networks architectures is proposed.

5.1.1.1 Non-optimized Trainings

In the non-optimized trainings, some parameters can be randomly established. In this book, four tests are performed in this granulation where mainly, the number of modules and the percentage of data for training have an important role. The best results obtained with the different biometric measures are presented below. In each test, 30 non-optimized trainings are performed.

© The Author(s) 2016
D. Sanchez and P. Melin, *Hierarchical Modular Granular Neural Networks with Fuzzy Aggregation*, SpringerBriefs in Computational Intelligence,
DOI 10.1007/978-3-319-28862-8_5

Test #1 (Using up to 80 % of Data for Training)

In this test, the number of modules and hidden layers are fixed where in each training 3 modules with 2 hidden layers are used. The percentage of data for the training phase using up to 80 % of the data, the data used for training, the number of persons in each module, and the number of neurons are randomly established. In Tables 5.1, 5.2, 5.3 and 5.4, the best 5 results of the face (CASIA Database), iris, ear and voice are respectively shown.

Table 5.1 The best 5 non-optimized results (with 3 modules, Face, 80 %)

Training	Images for training	Persons per module	Recognition rate (%)	Error
T1F1	78 % (1, 2, 3 and 5)	Module #1 (1 to 40) Module #2 (41 to 49) Module #3 (50 to 77)	71.43	0.2857
T1F4	72 % (2, 3, 4 and 5)	Module #1 (1 to 12) Module #2 (13 to 46) Module #3 (47 to 77)	89.61	0.1039
T1F18	48 % (4 and 5)	Module #1 (1 to 19) Module #2 (20 to 54) Module #3 (55 to 77)	71.43	0.2857
T1F24	36 % (1 and 5)	Module #1 (1 to 21) Module #2 (22 to 61) Module #3 (62 to 77)	72.29	0.2771
T1F25	81 % (1, 2, 4 and 5)	Module #1 (1 to 31) Module #2 (32 to 59) Module #3 (60 to 77)	77.92	0.2208

Table 5.2 The best 5 non-optimized Results (with 3 modules, Iris, 80 %)

Training	Images for training	Persons per module	Recognition rate (%)	Error
T1I6	80 % (1, 2, 3, 4, 5, 6, 7, 8, 9, 12 and 14)	Module #1 (1 to 36) Module #2 (37 to 45) Module #3 (46 to 77)	94.81	0.0519
T1I24	68 % (3, 4, 5, 6, 8, 9, 10, 11, 13 and 14)	Module #1 (1 to 26) Module #2 (27 to 60) Module #3 (61 to 77)	94.48	0.0552
T1I25	64 % (1, 3, 4, 5, 6, 10, 11, 13 and 14)	Module #1 (1 to 33) Module #2 (34 to 46) Module #3 (47 to 77)	94.55	0.0545
T1I29	77 % (1, 2, 3, 4, 5, 6, 7, 10, 12, 13 and 14)	Module #1 (1 to 30) Module #2 (31 to 52) Module #3 (53 to 77)	95.24	0.0476
T1I30	74 % (1, 2, 4, 5, 6, 7, 8, 10, 11 and 13)	Module #1 (1 to 30) Module #2 (31 to 47) Module #3 (48 to 77)	96.10	0.0390

Table 5.3 The best 5 non-optimized results (with 3 modules, Ear, 80 %)

Training	Images for training	Persons per module	Recognition rate (%)	Error
T1E2	70 % (2, 3 and 4)	Module #1 (1 to 22) Module #2 (23 to 50) Module #3 (51 to 77)	97.40	0.0260
T1E13	58 % (2 and 3)	Module #1 (1 to 31) Module #2 (32 to 48) Module #3 (49 to 77)	83.12	0.1688
T1E16	52 % (2 and 3)	Module #1 (1 to 10) Module #2 (11 to 39) Module #3 (40 to 77)	84.42	0.1558
T1E20	76 % (2, 3 and 4)	Module #1 (1 to 38) Module #2 (39 to 49) Module #3 (50 to 77)	97.40	0.0260
T1E26	74 % (2, 3 and 4)	Module #1 (1 to 39) Module #2 (40 to 56) Module #3 (57 to 77)	98.70	0.0130

Table 5.4 The best 5 non-optimized results (with 3 modules, Voice, 80 %)

Training	Images for training	Persons per module	Recognition rate (%)	Error
T1V13	66 % (1, 2, 3, 4, 8, 9 and 10)	Module #1 (1 to 31) Module #2 (32 to 68) Module #3 (69 to 77)	95.67	0.0433
T1V24	58 % (1, 6, 7, 8, 9 and 10)	Module #1 (1 to 9) Module #2 (10 to 54) Module #3 (55 to 77)	94.81	0.0519
T1V25	53 % (1, 2, 5, 6 and 9)	Module #1 (1 to 10) Module#2 (11 to 42) Module #3 (43 to 77)	93.25	0.0675
T1V26	55 % (1, 2, 5, 6, 7 and 9)	Module #1 (1 to 8) Module #2 (9 to 47) Module #3 (48 to 77)	93.83	0.0617
T1V28	50 % (1, 5, 7, 9 and 10)	Module #1 (1 to 22) Module #2 (23 to 41) Module #3 (42 to 77)	95.84	0.0416

Test #2 (Using up to 80 % of Data for Training)

In this test, the number of hidden layers is fixed; each module has 2 hidden layers. In each training, the number of modules is randomly established (from 1 to 10) and the percentage of data for the training phase using up to 80 % of the data, the data used for training, the number of persons in each module, and the number of neurons are also randomly established. In Tables 5.5, 5.6, 5.7 and 5.8, the best 5 results of the face (CASIA Database), iris, ear and voice are respectively shown.

Table 5.5 Non-optimized results (with different number of modules, Face, 80 %)

Training	Images for training	Persons per module	Recognition rate (%)	Error
T2F1	79 % (1, 3, 4 and 5)	Module #1 (1 to 10) Module #2 (11 to 34) Module #3 (35 to 39) Module #4 (40 to 61) Module #5 (62 to 71) Module #6 (72 to 77)	87.01	0.1299
T2F2	69 % (1, 4 and 5)	Module #1 (1 to 12) Module #2 (13 to 22) Module #3 (23 to 35) Module #4 (36 to 37) Module #5 (38 to 42) Module #6 (43 to 44) Module #7 (45 to 58) Module #8 (59 to 61) Module #9 (62 to 66) Module #10 (67 to 77)	85.71	0.1429
T2F14	51 % (1, 4 and 5)	Module #1 (1 to 25) Module #2 (26 to 47) Module #3 (48 to 55) Module #4 (56 to 77)	88.31	0.1169
T2F21	68 % (2, 3 and 5)	Module #1 (1 to 4) Module #2 (5 to 40) Module #3 (41 to 77)	80.52	0.1948
T2F28	54 % (3, 4 and 5)	Module #1 (15 to 29) Module #2 (30 to 77)	88.31	0.1169

In Tables 5.9 and 5.10, the results for the Essex Database are shown. These results are used as a basis to perform the granulation #2. In Table 5.9, the best result of the face (Essex Database) using the color RGB faces. In Table 5.10, the best results using type-1 fuzzy logic as edge detection are shown.

Test #3 (Using up to 50 % of Data for Training)

In this test, the number of modules and hidden layers are fixed, where in each training, 3 modules with 2 hidden layers are used. The percentage of data for the training phase using up to 50 % of the data, the data used for training, the number of persons in each module, and the number of neurons are randomly established. In Tables 5.11, 5.12, 5.13 and 5.14, the best 5 results of the face (CASIA Database), iris, ear and voice are respectively shown.

Test #4 (Using up to 50 % of Data for Training)

In this test, the number of hidden layers is fixed; each module has 2 hidden layers. In each training, the number of modules is randomly established (from 1 to 10) and

Table 5.6 The best 5 non-optimized results (with different number of modules, Iris, 80 %)

Training	Images for training	Persons per module	Recognition rate (%)	Error
T2I3	73 % (1, 2, 3, 5, 6, 8, 9, 10, 12 and 13)	Module #1 (1 to 24) Module #2 (25 to 77)	96.10	0.0390
T2I10	80 % (1, 2, 3, 5, 6, 9, 10, 11, 12, 13 and 14)	Module #1 (1 to 10) Module #2 (11 to 18) Module #3 (19 to 28) Module #4 (29 to 39) Module #5 (40 to 52) Module #6 (53 to 64) Module #7 (65 to 77)	98.27	0.0173
T2I12	75 % (1, 2, 3, 4, 5, 7, 8, 11, 12, 13 and 14)	Module #1 (1 to 8) Module #2 (9 to 24) Module #3 (25 to 28) Module #4 (29 to 32) Module #5 (33 to 44) Module #6 (45 to 55) Module #7 (56 to 69) Module #8 (70 to 77)	96.97	0.0303
T2I15	66 % (2, 3, 4, 5, 6, 8, 10, 11 and 13)	Module #1 (1 to 2) Module #2 (3 to 16) Module #3 (17 to 23) Module #4 (24 to 26) Module #5 (27 to 31) Module #6 (32 to 44) Module #7 (45 to 54) Module #8 (55 to 65) Module #9 (66 to 74) Module #10 (75 to 77)	97.66	0.0234
T2I23	76 % (1, 3, 4, 5, 6, 7, 8, 9, 10, 13 and 14)	Module #1 (1 to 10) Module #2 (11 to 26) Module #3 (27 to 31) Module #4 (32 to 47) Module #5 (48 to 60) Module #6 (61 to 77)	96.97	0.0303

the percentage of data for the training phase using up to 50 % of the data, the data used for training, the number of persons in each module, and the number of neurons are also randomly established. In Tables 5.15, 5.16, 5.17 and 5.18, the best results of the face (CASIA Database), iris, ear and voice are respectively shown.

5.1.1.2 Optimized Results

The results using the optimization described above for the modular granular neural networks are presented below. This optimization is divided into 2 tests; the first one using up to 80 % of data for training and the second one using up to 50 % of data for training, in each test 20 evolutions are performed.

Table 5.7 The best 5 non-optimized results (with different number of modules, Ear, 80 %)

Training	Images for training	Persons per module	Recognition rate (%)	Error
T2E1	74 % (1, 2 and 3)	Module #1 (1 to 5) Module #2 (6 to 15) Module #3 (16 to 26) Module #4 (27 to 30) Module #5 (31 to 43) Module #6 (44 to 49) Module #7 (50 to 58) Module #8 (59 to 71) Module #9 (72 to 77)	94.81	0.0519
T2E3	79 % (1, 2 and 3)	Module #1 (1 to 2) Module #2 (3 to 11) Module #3 (12 to 19) Module #4 (20 to 35) Module #5 (36 to 48) Module #6 (49 to 62) Module #7 (63 to 66) Module #8 (67 to 68) Module #9 (69 to 77)	96.10	0.0390
T2E5	81 % (2, 3 and 4)	Module #1 (1 to 12) Module #2 (13 to 14) Module #3 (15 to 23) Module #4 (24 to 43) Module #5 (44 to 58) Module #6 (59 to 77)	97.40	0.0260
T2E22	69 % (2, 3 and 4)	Module #1 (1 to 16) Module #2 (17 to 34) Module #3 (35 to 47) Module #4 (48 to 60) Module #5 (61 to 77)	96.10	0.0390
T2E28	77 % (2, 3 and 4)	Module #1 (1 to 5) Module #2 (6 to 54) Module #3 (55 to 77)	92.21	0.0779

Optimization up to 80 % of Data for Training

In Tables 5.19, 5.20, 5.21 and 5.22 and in Figs. 5.1, 5.2, 5.3 and 5.4, the best 5 results and the convergence of the best optimized result of the face (CASIA Database), iris, ear and voice are respectively shown.

Optimization up to 50 % of Data for Training

In Tables 5.23, 5.24, 5.25 and 5.26 and in Figs. 5.5, 5.6, 5.7 and 5.8, the best 5 results and the convergence of the best optimized result of the face (CASIA Database), iris, ear and voice are respectively shown.

Table 5.8 The best 5 non-optimized results (with different number of modules, Voice, 80 %)

Training	Images for training	Persons per module	Recognition rate (%)	Error
T2V6	64 % (1, 2, 3, 7, 8 and 9)	Module #1 (1 to 13) Module #2 (14 to 26) Module #3 (27 to 32) Module #4 (33 to 37) Module #5 (38 to 50) Module #6 (51 to 52) Module #7 (53 to 66) Module #8 (67 to 68) Module #9 (69 to 77)	93.18	0.0682
T2V18	73 % (1, 3, 4, 5, 6, 9 and 10)	Module #1 (1 to 3) Module #2 (4 to 13) Module #3 (14 to 23) Module #4 (24 to 35) Module #5 (36 to 41) Module #6 (42 to 53) Module #7 (54 to 62) Module #8 (63 to 73) Module #9 (74 to 77)	95.24	0.0476
T2V19	70 % (2, 3, 5, 6, 7, 9 and 10)	Module #1 (1 to 14) Module #2 (15 to 28) Module #3 (29 to 39) Module #4 (40 to 49) Module #5 (50 to 60) Module #6 (61 to 67) Module #7 (68 to 69) Module #8 (70 to 77)	96.10	0.0390
T2V21	70 % (2, 5, 6, 7, 8, 9 and 10)	Module #1 (1 to 9) Module #2 (10 to 13) Module #3 (14 to 16) Module #4 (17 to 22) Module #5 (23 to 33) Module #6 (34 to 45) Module #7 (46 to 54) Module #8 (55 to 64) Module #9 (65 to 75) Module #10 (76 to 77)	97.40	0.0260
T2V26	78 % (1, 3, 5, 6, 7, 8, 9 and 10)	Module #1 (1 to 31) Module #2 (32 to 77)	96.75	0.0325

5.1.2   Granulation #2

This second granulation is designed to be used with big databases and the main idea is depending on complexity level of the persons, the persons with the same complexity level are grouped. To prove the effectiveness of this method, 195 persons of the Essex Database are used.

Table 5.9 The best 5 non-optimized results (with different number of modules, Face, 80 %, color RGB)

Training	Images for training	Persons per module	Recognition rate (%)	Error
T2C1F1	48 % (2, 3, 9, 11, 12, 13, 15, 16, 18 and 19)	Module #1 (1 to 36) Module #2 (37 to 38) Module #3 (39 to 53) Module #4 (54 to 74) Module #5 (75 to 130) Module #6 (131 to 146) Module #7 (147 to 151) Module #8 (152 to 195)	74.72	0.2528
T2C1F6	30 % (1, 4, 6, 9, 14 and 16)	Module #1 (1 to 57) Module #2 (58 to 195)	89.38	0.1062
T2C1F17	66 % (1, 2, 3, 4, 6, 7, 8, 9, 10, 13, 15, 16 and 17)	Module #1 (1 to 12) Module #2 (13 to 29) Module #3 (30 to 54) Module #4 (55 to 77) Module #5 (78 to 82) Module #6 (83 to 105) Module #7 (106 to 123) Module #8 (124 to 148) Module #9 (149 to 168) Module #10 (169 to 195)	81.39	0.1861

Table 5.10 The best 5 non-optimized results (with different number of modules, Face, 80 %, Type-1 FIS)

Training	Images for training	Persons per module	Recognition rate (%)	Error
T2C2F3	37 % (2, 6, 9, 10, 13, 17 and 18)	Module #1 (1 to 11) Module #2 (12 to 29) Module #3 (30 to 58) Module #4 (59 to 91) Module #5 (92 to 112) Module #6 (113 to 131) Module #7 (132 to 141) Module #8 (142 to 175) Module #9 (176 to 195)	92.86	0.0714
T2C2F4	63 % (1, 3, 4, 10, 11, 12, 13, 14, 15, 16, 17, 18 and 19)	Module #1 (1 to 39) Module #2 (40 to 48) Module #3 (49 to 62) Module #4 (63 to 128) Module #5 (129 to 163) Module #6 (164 to 195)	95.46	0.0454
T2C2F6	42 % (6, 8, 10, 12, 14, 16, 17 and 20)	Module #1 (1 to 15) Module #2 (16 to 23) Module #3 (24 to 44) Module #4 (45 to 65) Module #5 (66 to 85) Module #6 (86 to 128) Module #7 (129 to 162) Module #8 (163 to 195)	94.66	0.0534

Table 5.11 The best 5 non-optimized results (with 3 modules, Face, 50 %)

Training	Images for training	Persons per module	Recognition rate (%)	Error
T3F8	43 % (4 and 5)	Module #1 (1 to 33) Module #2 (34 to 58) Module #3 (59 to 77)	68.40	0.3160
T3F11	38 % (2 and 3)	Module #1 (1 to 20) Module #2 (21 to 52) Module #3 (53 to 77)	66.67	0.3333
T3F20	43 % (3 and 5)	Module #1 (1 to 31) Module #2 (32 to 49) Module #3 (50 to 77)	73.59	0.2641
T3F21	48 % (1 and 5)	Module #1 (1 to 22) Module #2 (23 to 45) Module #3 (46 to 77)	69.26	0.3074
T3F26	34 % (2 and 4)	Module #1 (1 to 38) Module #2 (39 to 41) Module #3 (42 to 77)	67.10	0.3290

Table 5.12 The best 5 non-optimized results (with 3 modules, Iris, 50 %)

Training	Images for training	Persons per module	Recognition rate (%)	Error
T3I3	46 % (1, 2, 4, 9,12 and 13)	Module #1 (1 to 9) Module #2 (10 to 45) Module #3 (46 to 77)	89.29	0.1071
T3I5	42 % (1, 2, 7, 8, 13 and 14)	Module #1 (1 to 18) Module #2 (19 to 50) Module #3 (51 to 77)	89.45	0.1055
T3I19	28 % (7, 9, 11 and 13)	Module #1 (1 to 13) Module #2 (14 to 49) Module #3 (50 to 77)	88.70	0.1130
T3I22	41 % (1, 4, 5, 6, 10 and 14)	Module #1 (1 to 39) Module #2 (40 to 75) Module #3 (76 to 77)	89.45	0.1055
T3I27	33 % (2, 4, 5, 11 and 13)	Module #1 (1 to 25) Module #2 (26 to 38) Module #3 (39 to 77)	89.03	0.1097

5.1.2.1 Previous Analysis

As in Sect. 3.1.2 is described, firstly an analysis is performed. A comparison between the averages and the best results of the non-optimized trainings are presented in Table 5.27.

The analysis already described is used. The analysis is performed with the second pre-processing, because a best average is obtained with this one. The percentage of complexity (*poc*) for each person is calculated. The rules used to perform the granulation are shown in Fig. 5.9.

Table 5.13 The best 5 non-optimized results (with 3 modules, Ear, 50 %)

Training	Images for training	Persons per module	Recognition rate (%)	Error
T3E7	45 % (1 and 2)	Module #1 (1 to 25) Module #2 (26 to 59) Module #3 (60 to 77)	77.92	0.2208
T3E9	45 % (1 and 3)	Module #1 (1 to 25) Module #2 (26 to 51) Module #3 (52 to 77)	85.06	0.1494
T3E11	39 % (2 and 4)	Module #1 (1 to 13) Module #2 (14 to 51) Module #3 (52 to 77)	70.13	0.2987
T3E13	48 % (1 and 2)	Module #1 (1 to 26) Module #2 (27 to 39) Module #3 (40 to 77)	70.78	0.2922
T3E29	42 % (2 and 3)	Module #1 (1 to 18) Module #2 (19 to 37) Module #3 (38 to 77)	87.66	0.1234

Table 5.14 The best 5 non-optimized results (with 3 modules, Voice, 50 %)

Training	Images for training	Persons per module	Recognition rate (%)	Error
T3V18	46 % (1, 2, 4, 7 and 9)	Module #1 (1 to 18) Module #2 (19 to 66) Module #3 (67 to 77)	91.43	0.0857
T3V24	37 % (1, 6, 7 and 10)	Module #1 (1 to 9) Module #2 (10 to 54) Module #3 (55 to 77)	95.24	0.0476
T3V25	34 % (1, 6 and 9)	Module #1 (1 to 10) Module #2 (11 to 42) Module #3 (43 to 77)	90.21	0.0909
T3V26	35 % (1, 2, 5 and 7)	Module #1 (1 to 8) Module #2 (9 to 47) Module #3 (48 to 77)	90.48	0.0952
T3V28	32 % (5, 7 and 9)	Module #1 (1 to 22) Module #2 (23 to 41) Module #3 (42 to 77)	88.13	0.1187

The number of persons with the same complexity level is shown in Table 5.28.

5.1.2.2   Optimized Results

With this granulation, 5 evolutions are performed. The results obtained in the best evolution are shown below. In this case the best evolution is the #2. In Table 5.29, the randomly selected validations that were used in Evolution #2 are shown. The images used for the training and the testing phases are also shown in Table 5.29.

Table 5.15 The best 5 non-optimized results (with different number of modules, Face, 50 %)

Training	Images for training	Persons per module	Recognition rate (%)	Error
T4F2	34 % (3 and 5)	Module #1 (1 to 13) Module #2 (14 to 22) Module #3 (23 to 42) Module #4 (43 to 50) Module #5 (51 to 54) Module #6 (55 to 74) Module #7 (75 to 77)	77.49	0.2251
T4F18	51 % (3, 4 and 5)	Module #1 (1 to 26) Module #2 (27 to 46) Module #3 (47 to 77)	89.61	0.1039
T4F23	50 % (3, 4 and 5)	Module #1 (1 to 2) Module #2 (3 to 12) Module #3 (13 to 18) Module #4 (19 to 29) Module #5 (30 to 34) Module #6 (35 to 36) Module #7 (37 to 48) Module #8 (49 to 53) Module #9 (54 to 65) Module #10 (66 to 77)	91.56	0.0844
T4F24	40 % (3 and 5)	Module #1 (1 to 13) Module #2 (14 to 18) Module #3 (19 to 34) Module #4 (35 to 44) Module #5 (45 to 55) Module #6 (56 to 62) Module #7 (63 to 66) Module #8 (67 to 72) Module #9 (73 to 77)	83.98	0.1602
T4F27	41 % (3 and 5)	Module #1 (1 to 13) Module #2 (14 to 27) Module #3 (28 to 39) Module #4 (40 to 44) Module #5 (45 to 57) Module #6 (58 to 61) Module #7 (62 to 73) Module #8 (74 to 77)	77.92	0.2208

The best errors and the averages in the external memory are shown in Table 5.30. In this Table, the deactivation of granules and validations can be observed. In this evolution a 0.0318 of recognition error was achieved, which means a 96.82 of recognition rate. In Table 5.31, the number of generations where the granule or validations were deactivated are shown.

The behavior of evolution #2 can be noticed in Fig. 5.10, where the best, the average and the worst results obtained in each generation are shown.

In Table 5.32, the architecture of the modular granular neural network and the results obtained for each validation are shown.

Table 5.16 The best 5 non-optimized results (with different number of modules, Face, 50 %)

Training	Images for training	Persons per module	Recognition rate (%)	Error
T4I3	41 % (4, 6, 9, 11, 13 and 14)	Module #1 (1 to 11) Module #2 (12 to 21) Module #3 (22 to 27) Module #4 (28 to 35) Module #5 (36 to 37) Module #6 (38 to 49) Module #7 (50 to 55) Module #8 (56 to 61) Module #9 (62 to 77)	93.18	0.0682
T4I6	45 % (3, 4, 7, 8, 11 and 13)	Module #1 (1 to 2) Module #2 (3 to 10) Module #3 (11 to 16) Module #4 (17 to 32) Module #5 (33 to 37) Module #6 (38 to 48) Module #7 (49 to 60) Module #8 (61 to 77)	90.75	0.0925
T4I17	47 % (2, 4, 6, 7, 9, 11 and 14)	Module #1 (1 to 11) Module #2 (12 to 20) Module #3 (21 to 31) Module #4 (32 to 36) Module #5 (37 to 38) Module #6 (39 to 47) Module #7 (48 to 55) Module #8 (56 to 66) Module #9 (67 to 77)	94.25	0.0575
T4I19	51 % (1, 5, 6, 7, 8, 10 and 13)	Module #1 (1 to 12) Module #2 (13 to 19) Module #3 (20 to 21) Module #4 (22 to 30) Module #5 (31 to 40) Module #6 (41 to 49) Module #7 (50 to 61) Module #8 (62 to 64) Module #9 (65 to 77)	94.25	0.0575
T4I20	48 % (2, 3, 4, 7, 11, 13 and 14)	Module #1 (1 to 14) Module #2 (15 to 20) Module #3 (21 to 31) Module #4 (32 to 34) Module #5 (35 to 44) Module #6 (45 to 54) Module #7 (55 to 64) Module #8 (65 to 70) Module #9 (71 to 77)	95.55	0.0445

In Fig. 5.11, the general average stored in the external memory in each generation is shown. In Fig. 5.12, the average stored in the external memory for each granule is shown.

Table 5.17 The best 5 non-optimized results (with different number of modules, Ear, 50 %)

Training	Images for training	Persons per module	Recognition rate (%)	Error
T4E8	45 % (2 and 4)	Module #1 (1 to 6) Module #2 (7 to 12) Module #3 (13 to 34) Module #4 (35 to 43) Module #5 (44 to 69) Module #6 (70 to 77)	79.87	0.2013
T4E14	47 % (1 and 3)	Module #1 (1 to 12) Module #2 (13 to 26) Module #3 (27 to 38) Module #4 (39 to 44) Module #5 (45 to 53) Module #6 (54 to 66) Module #7 (67 to 77)	77.27	0.2273
T4E23	48 % (3 and 4)	Module #1 (1 to 18) Module #2 (19 to 25) Module #3 (26 to 43) Module #4 (44 to 63) Module #5 (64 to 77)	78.57	0.2143
T4E26	36 % (1)	Module #1 (1 to 10) Module #2 (11 to 24) Module #3 (25 to 32) Module #4 (33 to 42) Module #5 (43 to 55) Module #6 (56 to 65) Module #7 (66 to 77)	73.59	0.2641
T4E30	47 % (2 and 3)	Module #1 (1 to 17) Module #2 (18 to 27) Module #3 (28 to 46) Module #4 (47 to 77)	83.12	0.1688

Table 5.18 The best 5 non-optimized results (with different number of modules, Voice, 50 %)

Training	Images for training	Persons per module	Recognition rate (%)	Error
T4V6	45 % (1, 5, 6, 9 and 10)	Module #1 (1 to 5) Module #2 (6 to 16) Module #3 (17 to 26) Module #4 (27 to 33) Module #5 (34 to 49) Module #6 (50 to 58) Module #7 (59 to 63) Module #8 (64 to 77)	96.10	0.0390
T4V7	48 % (1, 3, 4, 5 and 10)	Module #1 (1 to 9) Module #2 (10 to 18) Module #3 (19 to 25) Module #4 (26 to 37) Module #5 (38 to 48) Module #6 (49 to 60) Module #7 (61 to 71) Module #8 (72 to 77)	96.62	0.0338

(continued)

Table 5.18 (continued)

Training	Images for training	Persons per module	Recognition rate (%)	Error
T4V20	49 % (2, 4, 8, 9 and 10)	Module #1 (1 to 8) Module #2 (9 to 21) Module #3 (22 to 41) Module #4 (42 to 43) Module #5 (44 to 55) Module #6 (56 to 69) Module #7 (70 to 77)	95.06	0.0494
T4V22	42 % (1, 3, 7 and 8)	Module #1 (1 to 14) Module #2 (15 to 28) Module #3 (29 to 31) Module #4 (32 to 44) Module #5 (45 to 60) Module #6 (61 to 64) Module #7 (65 to 69) Module #8 (70 to 72) Module #9 (73 to 77)	92.86	0.0714
T4V23	45 % (1, 2, 4, 6 and 8)	Module #1 (1 to 7) Module #2 (8 to 13) Module #3 (14 to 18) Module #4 (19 to 27) Module #5 (28 to 38) Module #6 (39 to 44) Module #7 (45 to 52) Module #8 (53 to 64) Module #9 (65 to 77)	93.77	0.0623

Table 5.19 The best 5 optimized results (Face, 80 %)

Evolution	Images for training	Num. hidden layers and Num. of neurons	Persons per module	Rec. rate (%)	Error
OP1F4	50 % (3, 4 and 5)	*1 (32)* *1 (149)* *1 (205)* *1 (193)* *1 (292)* *1 (147)* *1 (130)* *1 (283)* *1 (263)* *1 (290)*	Module #1 (1 to 9) Module #2 (10 to 16) Module #3 (17 to 26) Module #4 (27 to 34) Module #5 (35 to 37) Module #6 (38 to 52) Module #7 (53 to 59) Module #8 (60 to 65) Module #9 (66 to 73) Module #10 (74 to 77)	97.40	0.0260
OP1F9	50 % (3, 4 and 5)	*4 (29, 68, 83, 222)* *12 (108, 46)* *4 (235, 68, 147, 88)* *1 (256)* *1 (269)* *1 (286)* *2 (178, 25)* *2 (296, 242)*	Module #1 (1 to 2) Module #2 (3 to 18) Module #3 (19 to 34) Module #4 (35 to 49) Module #5 (50 to 51) Module #6 (52 to 66) Module #7 (67 to 71) Module #8 (72 to 77)	97.40	0.0260

(continued)

Table 5.19 (continued)

Evolution	Images for training	Num. hidden layers and Num. of neurons	Persons per module	Rec. rate (%)	Error
OP1F16	50 % (3, 4 and 5)	1 (197) 1 (164) 1 (272) 1 (196) 1 (20) 1 (188)	Module #1 (1 to 27) Module #2 (28 to 40) Module #3 (41 to 66) Module #4 (67 to 72) Module #5 (73 to 75) Module #6 (76 to 77)	98.05	0.0195
OP1F17	50 % (3, 4 and 5)	2 (200, 199) 5 (267, 170, 282, 214, 88) 4 (228, 88, 193, 120) 1 (300) 3 (297, 152, 120) 1 (21) 4 (134, 238, 41, 274) 1 (51) 2 (297, 209) 5 (292, 118, 58, 225, 202)	Module #1 (1 4) Module #2 (5 to 8) Module #3 (9 to 22) Module #4 (23 to 38) Module #5 (39 to 47) Module #6 (48 to 49) Module #7 (50 to 56) Module #8 (57 to 67) Module #9 (68 to 71) Module #10 (72 to 77)	97.40	0.0260
OP1F20	50 % (1, 4 and 5)	1 (248) 1 (192) 1 (98) 1 (50) 1 (75) 1 (125) 1 (69) 1 (153) 1 (163) 1 (73)	Module #1 (1 to 14) Module #2 (15 to 16) Module #3 (17 to 18) Module #4 (19 to 26) Module #5 (27 to 42) Module #6 (43 to 44) Module #7 (45 to 57) Module #8 (58 to 67) Module #9 (68 to 75) Module #10 (76 to 77)	97.40	0.0260

Table 5.20 The best 5 optimized results (Iris, 80 %)

Evolution	Images for training	Num. hidden layers and Num. of neurons	Persons per module	Rec. rate (%)	Error
OP1I4	69 % (1, 2, 3, 4, 5, 6, 8, 12, 13 and 14)	3 (287, 36, 155) 4 (297, 184, 251, 26) 3 (225, 31, 23) 1 (162) 3 (93, 118, 34) 2 (157, 181) 4 (163, 286, 145, 85) 3 (87, 50, 167) 1 (60)	Module #1 (1 to 3) Module #2 (4 to 15) Module #3 (16 to 29) Module #4 (30 to 42) Module #5 (43 to 47) Module #6 (48 to 49) Module #7 (50 to 56) Module #8 (57 to 61) Module #9 (62 to 77)	99.35	0.0065
OP1I11	79 % (1, 2, 3, 4, 5, 6, 7, 8, 10, 11 and 14)	1 (245) 1 (130) 1 (272) 1 (171) 1 (211) 1 (280)	Module #1 (1 to 6) Module #2 (7 to 20) Module #3 (21 to 24) Module #4 (25 to 38) Module #5 (39 to 58) Module #6 (59 to 77)	99.57	0.0043

(continued)

Table 5.20 (continued)

Evolution	Images for training	Num. hidden layers and Num. of neurons	Persons per module	Rec. rate (%)	Error
OP1I13	71 % (1, 2, 3, 5, 6, 8, 9, 11, 13 and 14)	*1 (168)* *2 (120, 164)* *2 (286, 232)* *2 (78, 134)* *2 (296, 293)* *4 (152, 33, 216, 160)* *4 (184, 55, 64, 48)* *2 (114, 240)* *2 (230, 27)*	Module #1 (1 to 14) Module #2 (15 to 20) Module #3 (21 to 22) Module #4 (23 to 30) Module #5 (31 to 36) Module #6 (37 to 40) Module #7 (41 to 46) Module #8 (47 to 62) Module #9 (63 to 77)	99.68	0.0032
OP1I16	79 % (1, 2, 3, 5, 6, 7, 8, 11, 12, 13 and 14)	*1 (209)* *1 (133)* *1 (238)* *1 (203)* *1 (118)* *1 (117)* *1 (147)* *1 (46)* *1 (67)*	Module #1 (1 to 5) Module #2 (6 to 10) Module #3 (11 to 22) Module #4 (23 to 32) Module #5 (33 to 36) Module #6 (37 to 46) Module #7 (47 to 55) Module #8 (56 to 71) Module #9 (72 to 77)	99.57	0.0043
OP1I19	75 % (1, 3, 4, 5, 6, 8, 9, 10, 11, 13 and 14)	*1 (75)* *1 (300)* *1 (211)* *1 (257)* *1 (234)* *1 (139)*	Module #1 (1 to 11) Module #2 (12 to 24) Module #3 (25 to 37) Module #4 (38 to 46) Module #5 (47 to 60) Module #6 (61 to 77)	99.57	0.0043

Table 5.21 The best 5 optimized results (Ear, 80 %)

Evolution	Images for training	Num. hidden layers and Num. of neurons	Persons per module	Rec. rate (%)	Error
OP1E1	79 % (2, 3 and 4)	*4 (62,104, 29,127)* *3 (69, 51, 34)* *4 (96, 95, 92, 98)* *2 (38, 149)* *4 (45, 83,3 2, 153)* *5 (156, 127, 38, 29, 117)* *3 (125, 157, 30)* *3 (43, 100, 168)*	Module #1 (1 to 8) Module #2 (9 to 10) Module # 3 (11 to 23) Module #4 (24 to 31) Module #5 (32 to 47) Module #6 (48 to 57) Module #7 (58 to 62) Module #8 (63 to 77)	77/77 100	0
OP1E2	66 % (1, 2 and 3)	*1 (103)* *4 (184, 139, 125, 188)* *2 (28,129)* *1 (131)* *5 (91, 139, 50, 76, 190)* *4 (56, 152, 165, 148)* *4 (108, 192, 191, 137)* *4 (73, 171, 49, 184)*	Module #1 (1 to 2) Module #2 (3 to 11) Module #3 (12 to 20) Module #4 (21 to 33) Module #5 (34 to 49) Module #6 (50 to 54) Module #7 (55 to 66) Module #8 (67 to 77)	77/77 100	0

(continued)

Table 5.21 (continued)

Evolution	Images for training	Num. hidden layers and Num. of neurons	Persons per module	Rec. rate (%)	Error
OP1E3	72 % (2, 3 and 4)	3 (165, 197, 153) 5 (97, 68, 67, 105, 168) 1 (68) 4 (72, 190, 83, 68) 2 (157, 115) 4 (141, 48, 45, 145) 2 (140, 124) 3 (46, 188, 74) 4 (114, 153, 30, 80)	Module #1 (1 to 5) Module #2 (6 to 22) Module #3 (23 to 26) Module #4 (27 to 43) Module #5 (44 to 52) Module #6 (53 to 55) Module #7 (56 to 57) Module #8 (58 to 72) Module #9 (73 to 77)	77/77 100	0
OP1E4	70 % (2, 3 and 4)	2 (156, 193) 2 (159, 96) 2 (21, 112) 3 (165, 129, 21)	Module #1(1 to14) Module #2 (15 to 36) Module #3 (37 to 55) Module #4 (56 to 77)	77/77 100	0
OP1E6	80 % (2, 3 and 4)	5 (90, 98, 185, 199, 168) 4 (54, 154, 167, 37)	Module #1 (1 to 24) Module #2 (25 to 77)	77/77 100	0

Table 5.22 The best 5 results optimized (Voice, 80 %)

Evolution	Images for training	Num. hidden layers and Num. of neurons	Persons per module	Rec. rate (%)	Error
OP1V4	75 % (1, 2, 3, 4, 5, 8, 9 and 10)	3 (55, 224, 223) 2 (68, 178) 2 (51, 85) 2 (114, 214) 3 (154, 295, 246)	Module #1 (1 to 14) Module #2 (15 to 32) Module #3 (33 to 36) Module #4 (37 to 62) Module #5 (63 to 77)	100	0
OP1V5	79 % (1, 2, 5, 6, 7, 8, 9 and 10)	2 (99, 224) 2 (183, 188) 2 (261, 57) 4 (100, 127, 293, 281) 1 (135) 5 (212, 281, 108, 35, 283) 5 (163, 21, 291, 74, 255) 1 (182)	Module #1 (1 to 8) Module #2 (9 to 11) Module #3 (12 to 24) Module #4 (25 to 35) Module #5 (36 to 49) Module #6 (50 to 59) Module #7 (60 to 67) Module #8 (68 to 77)	100	0
OP1V7	73 % (1, 3, 5, 7, 8, 9 and 10)	1 (150) 1 (198) 1 (160) 1 (242) 1 (287) 1 (245) 1 (61) 1 (36) 1 (194)	Module #1 (1 to 5) Module #2 (6 to 15) Module #3 (16 to 23) Module #4 (24 to 38) Module #5 (39 to 46) Module #6 (47 to 58) Module #7 (59 to 73) Module #8 (74 to 75) Module #9 (76 to 77)	100	0
OP1V8	78 % (1, 2, 4, 5, 6, 8, 9 and 10)	1 (227) 1 (47) 1 (144)	Module #1 (1 to 24) Module #2 (25 to 59) Module #3 (60 to 77)	100	0

(continued)

Table 5.22   (continued)

Evolution	Images for training	Num. hidden layers and Num. of neurons	Persons per module	Rec. rate (%)	Error
OP1V9	77 % (1, 2, 3, 6, 7, 8, 9 and 10)	4 (43, 199, 291, 270) 4 (158, 184, 120, 291) 2 (78, 40) 2 (247, 131) 1 (51) 3 (276, 205, 106) 1 (275) 4 (148, 256, 211, 57) 4 (62, 274, 163, 296) 1 (66, 93)	Module #1 (1 to 10) Module #2 (11 to 17) Module #3 (18 to 29) Module #4 (30 to 33) Module #5 (34 to 47) Module #6 (48 to 52) Module #7 (53 to 58) Module #8 (59 to 60) Module #9 (61 to 73) Module #10 (74 to 77)	100	0

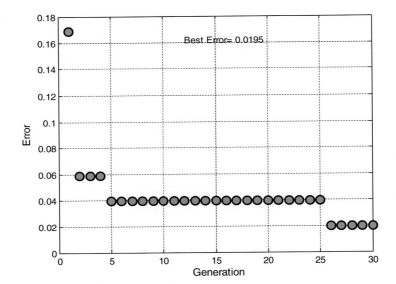

Fig. 5.1   Convergence of the evolution OP1F16 (Face)

The average stored in the external memory for each validation is shown in Fig. 5.13 and a zoom is presented in Fig. 5.14.

5.2   Fuzzy Integration Results

The responses combination of the modular granular neural networks is performed using fuzzy logic as an integration technique. Six cases are established for combining different trainings of face (CASIA Database), iris, ear and voice of non-optimized and optimized results. In Table 5.33, the biometric measure results used in the different cases and their combination can be observed.

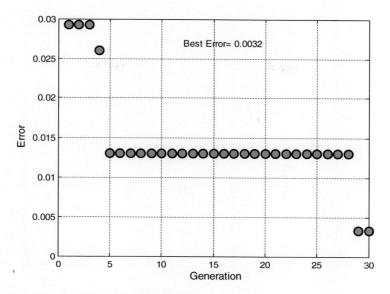

Fig. 5.2 Convergence of the evolution OP1I13 (Iris)

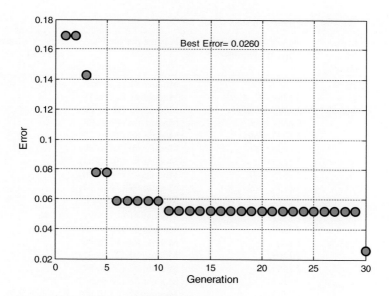

Fig. 5.3 Convergence of the evolution OP1E1 (Ear)

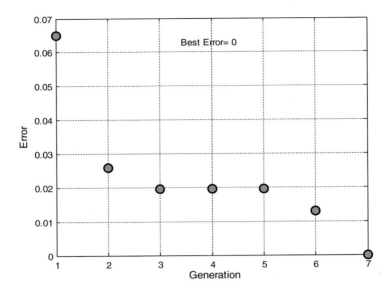

Fig. 5.4 Convergence of the evolution OP1V4 (Voice)

Table 5.23 The best 5 optimized results (Face, 50 %)

Evolution	Images for training	Num. hidden layers and Num. of neurons	Persons per module	Rec. rate (%)	Error
OP2F1	50 % (3, 4 and 5)	*2 (215, 209)* *3 (269, 92, 290)* *3 (24, 36, 274)* *5 (123, 133, 53, 73, 125)* *1 (233, 269)* *2 (100, 46)* *1 (125)* *2 (51, 199)* *3 (136, 139, 230)* *4 (171, 205, 291, 140)*	Module #1 (1 to 5) Module #2 (6 to 15) Module #3 (16 to 18) Module #4 (19 to 32) Module #5 (33 to 37) Module #6 (38 to 49) Module #7 (50 to 63) Module #8 (64 to 70) Module #9 (71 to 74) Module #10 (75 to 77)	96.75	0.0325
OP2F4	50 % (3, 4 and 5)	*1 (32)* *1 (149)* *1 (205)* *1 (193)* *1 (292)* *1 (147)* *1 (130)* *1 (283)* *1 (263)* *1 (290)*	Module #1 (1 to 9) Module #2 (10 to 16) Module #3 (17 to 26) Module #4 (27 to 34) Module #5 (35 to 37) Module #6 (38 to 52) Module #7 (53 to 59) Module #8 (60 to 65) Module #9 (66 to 73) Module #10 (74 to 77)	97.40	0.0260

(continued)

Table 5.23 (continued)

Evolution	Images for training	Num. hidden layers and Num. of neurons	Persons per module	Rec. rate (%)	Error
OP2F6	50 % (3, 4 and 5)	*1 (148)* *1 (283)* *1 (23)* *1 (83)* *1 (52)*	Module #1 (1 to 18) Module #2 (19 to 44) Module #3 (45 to 49) Module #4 (50 to 70) Module #5 (71 to 77)	96.75	0.0325
OP2F7	50 % (3, 4 and 5)	*1 (257)* *1 (268)* *1 (210)* *1 (286)* *1 (244)* *1 (82)* *1 (92)*	Module #1 (1 to 7) Module #2 (8 to 24) Module #3 (25 to 39) Module #4 (40 to 41) Module #5 (42 to 57) Module #6 (58 to 71) Module #7 (72 to 77)	96.75	0.0325
OP2F9	50 % (3, 4 and 5)	*4 (29, 68, 83, 222)* *2 (108, 46)* *4 (235, 68, 147, 88)* *1 (256)* *1 (269)* *1 (286)* *2 (178, 25)* *2 (296, 242)*	Module #1 (1 to 2) Module #2 (3 to 18) Module #3 (19 to 34) Module #4 (35 to 49) Module #5 (50 to 51) Module #6 (52 to 66) Module #7 (67 to 71) Module #8 (72 to 77)	97.40	0.0260

Table 5.24 The best 5 optimized results (Iris, 50 %)

Evolution	Images for training	Num. hidden layers and Num. of neurons	Persons per module	Rec. rate (%)	Error
OP2I4	48 % (2, 3, 4, 5, 7, 13 and 14)	*1 (134)* *1 (157)* *1 (188)* *1 (275)* *1 (230)* *1 (231)* *1 (108)* *1 (25)*	Module #1 (1 to 15) Module #2 (16 to 17) Module #3 (18 to 30) Module #4 (31 to 41) Module #5 (42 to 44) Module #6 (45 to 57) Module #7 (58 to 62) Module #8 (63 to 77)	97.40	0.0260
OP2I5	50 % (2, 3, 5, 6, 11, 13 and 14)	*1 (52)* *1 (145)* *1 (202)* *1 (105)* *1 (223)* *1 (63)* *1 (226)* *1 (229)* *1 (78)*	Module #1 (1 to 10) Module #2 (11 to 13) Module #3 (14 to 17) Module #4 (18 to 24) Module #5 (25 to 39) Module #6 (40 to 45) Module #7 (46 to 60) Module #8 (61 to 67) Module #9 (68 to 77)	97.40	0.0260

(continued)

Table 5.24 (continued)

Evolution	Images for training	Num. hidden layers and Num. of neurons	Persons per module	Rec. rate (%)	Error
OP2I10	46 % (2, 3, 5, 6, 7 and 13)	1 (122) 1 (218) 1 (49) 1 (122) 1 (233) 1 (97) 1 (204) 1 (267) 1 (53) 1 (176)	Module #1 (1 to 10) Module #2 (11 to 26) Module #3 (27 to 32) Module #4 (33 to 36) Module #5 (37 to 47) Module #6 (48 to 50) Module #7 (51 to 58) Module #8 (59 to 73) Module #9 (74 to 75) Module #10 (76 to 77)	97.24	0.0276
OP2I16	47 % (1, 3, 5, 6, 8, 9 and 13)	1 (151) 1 (48) 1 (94) 1 (97) 1 (158) 1 (149) 1 (176)	Module #1 (1 to 3) Module #2 (4 to 18) Module #3 (19 to 35) Module #4 (36 to 51) Module #5 (52 to 61) Module #6 (62 to 72) Module #7 (73 to 77)	97.40	0.0260
OP2I19	48 % (3, 4, 5, 6, 8, 9 and 11)	1 (66) 1 (179) 1 (218) 1 (105) 1 (249) 1 (159) 1 (245) 1 (153) 1 (132) 1 (163)	Module #1 (1 to 6) Module #2 (7 to 11) Module #3 (12 to 21) Module #4 (22 to 28) Module #5 (29 to 31) Module #6 (32 to 37) Module #7 (38 to 42) Module #8 (43 to 48) Module #9 (49 to 61) Module #10 (62 to 77)	97.77	0.0223

Table 5.25 The best 5 results optimized (Ear, 50 %)

Evolution	Images for training	Num. hidden layers and Num. of neurons	Persons per module	Rec. rate	Error
OP2E3	41 % (2 and 3)	5 (95, 56, 190, 43, 197) 3 (61, 82, 76) 2 (48, 63) 4 (36, 58, 142, 77) 2 (133, 175) 2 (141, 48)	Module #1 (1 to 7) Module #2 (8 to 21) Module #3 (22 to 32) Module #4 (33 to 42) Module #5 (43 to 53) Module #6 (54 to 77)	97.40 % (150/154)	0.0260
OP2E4	50 % (2 and 4)	5 (95, 56, 200, 43, 83) 3 (129, 82, 76) 2 (138, 142) 4 (118, 41, 200, 136) 2 (200, 142) 4 (147, 161, 47, 20)	Module #1 (1 to 15) Module #2 (16 to 30) Module #3 (31 to 38) Module #4 (39 to 54) Module #5 (55 to 65) Module #6 (66 to 77)	96.10 % (148/154)	0.0390

(continued)

Table 5.25 (continued)

Evolution	Images for training	Num. hidden layers and Num. of neurons	Persons per module	Rec. rate	Error
OP2E5	49 % (2 and 3)	*2 (69, 76)* *4 (170, 127, 29, 103)* *3 (126, 94, 125)* *1 (196)* *2 (21, 190)* *4 (178, 89, 86, 58)* *2 (190, 22)* *1 (74)* *2 (130, 106)* *5 (57, 169, 56, 101, 89)*	Module #1 (1 to 6) Module #2 (7 to 14) Module #3 (15 to 20) Module #4 (21 to 24) Module #5 (25 to 33) Module #6 (34 to 42) Module #7 (43 to 53) Module #8 (54 to 72) Module #9 (73 to 75) Module #10 (76 to 77)	96.10 % (148/154)	0.0390
OP2E8	41 % (2 and 3)	*5 (95, 56, 190, 43, 197)* *5 (61, 82, 76, 26, 79)* *1 (115)* *3 (118, 41, 170)* *2 (195, 47)* *2 (50, 169)*	Module #1 (1 to 10) Module #2 (11 to 19) Module #3 (20 to 37) Module #4 (38 to 54) Module #5 (55 to 73) Module #6 (74 to 77)	96.10 % (148/154)	0.0390
OP2E16	48 % (2 and 3)	*2 (167, 144)* *2 (85, 153)* *5 (90, 182, 200, 102, 126)* *1 (85)* *4 (32, 181, 116, 185)* *3 (27, 153, 97)* *3 (89, 140, 148)* *3 (126, 74, 67)*	Module #1 (1 to 19) Module #2 (20 to 33) Module #3 (34 to 40) Module #4 (41 to 56) Module #5 (57 to 66) Module #6 (67 to 69) Module #7 (70 to 71) Module #8 (72 to 77)	96.75 %	0.0325

Table 5.26 The best 5 results optimized (Voice, 50 %)

Evolution	Images for training	Num. Hidden layers and Num. of neurons	Persons per module	Rec. rate (%)	Error
OP1V4	42 % (2, 7, 9 and 10)	*2 (86, 102)* *1 (57)* *1 (227)* *1 (58)* *1 (111)* *4 (63, 83, 159, 89)* *4 (84, 209, 33, 187)*	Module #1 (1 to 9) Module #2 (10 to 24) Module #3 (25 to 34) Module #4 (35 to 48) Module #5 (49 to 57) Module #6 (58 to 61) Module #7 (62 to 77)	98.27	0.0173
OP1V6	48 % (2, 5, 8, 9 and 10)	*2 (157, 32)* *3 (273, 53, 225)* *1 (291)* *2 (130, 93)* *3 (247, 99, 210)* *3 (244, 156, 296)*	Module #1 (1 to 6) Module #2 (7 to 24) Module #3 (25 to 43) Module #4 (44 to 56) Module #5 (57 to 71) Module #6 (72 to 77)	98.18	0.0182

Table 5.26 (continued)

Evolution	Images for training	Num. Hidden layers and Num. of neurons	Persons per module	Rec. rate (%)	Error
OP1V7	47 % (1, 2, 5, 9 and 10)	*1 (146)* *1 (276)* *1 (296)* *1 (163)* *1 (202)* *1 (127)* *1 (249)*	Module #1 (1 to 10) Module #2 (11 to 26) Module #3 (27 to 40) Module #4 (41 to 56) Module #5 (57 to 61) Module #6 (62 to 74) Module #7 (75 to 77)	98.96	0.0182
OP1V16	49 % (1, 3, 8, 9 and 10)	*1 (188)* *1 (124)* *1 (253)* *1 (153)* *1 (198)* *1 (247)* *1 (159)* *1 (282)* *1 (203)*	Module #1 (1 to 7) Module #2 (8 to 11) Module #3 (12 to 26) Module #4 (27 to 37) Module #5 (38 to 42) Module #6 (43 to 49) Module #7 (50 to 64) Module #8 (65 to 68) Module #9 (69 to 77)	98.96	0.0104
OP1V18	48 % (1, 2, 6, 7 and 10)	*1 (129)* *2 (215, 37)* *5 (54, 144, 172, 105, 43)* *4 (273, 214, 298, 237)* *3 (175, 136, 168)* *2 (153, 263)* *1 (282)*	Module #1 (1 to 16) Module #2 (17 to 28) Module #3 (29 to 42) Module #4 (43 to 48) Module #5 (49 to 61) Module #6 (62 to 66) Module #7 (67 to 77)	98.96	0.0104

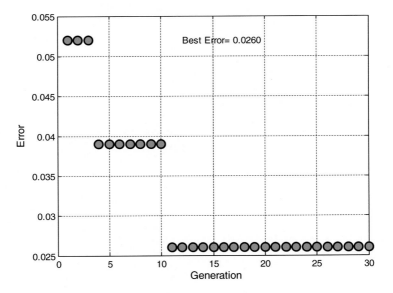

Fig. 5.5 Convergence of the evolution OP2F4 (Face)

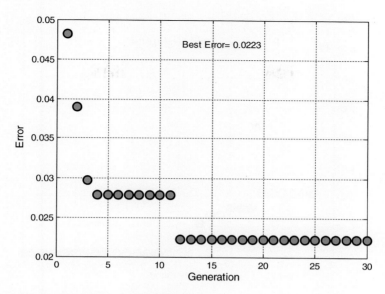

Fig. 5.6 Convergence of the evolution OP2I19 (Iris)

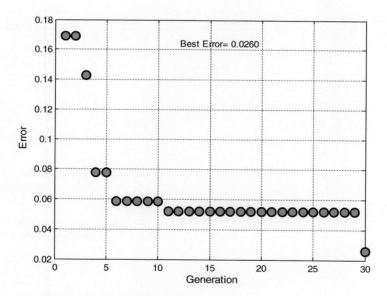

Fig. 5.7 Convergence of the evolution OP2E3 (Ear)

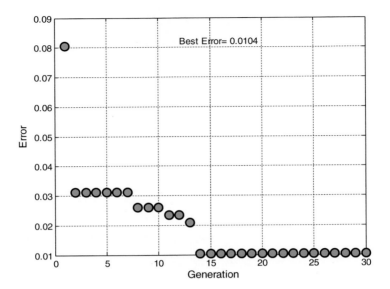

Fig. 5.8 Convergence of the evolution OP2V7 (Voice)

Table 5.27 Comparison of Essex database (non-optimized)

Preprocessing	Best	Average
Color RGB	89.38 % 0.1062	46.82 % 0.5318
Type-1 Fuzzy Logic	95.46 % 0.0454	53.21 % 0.4679

Fig. 5.9 Rules for the Essex database

If *poc* > 55 then
                              Complexity Level = High
Else

If *poc* > 45 and *poc* <= 55 then
                              Complexity Level = Medium
Else

If *poc* >= 0 and *poc* <= 45 then
                              Complexity Level = Low

End

Table 5.28 Granulation non-optimized (Essex database)

Granule	Number of persons	Complexity level
1	68	High
2	65	Medium
3	62	Low

Table 5.29 Validations used in evolution #2 (Essex database)

Validation	Percentage/Total		Images	
	Training	Testing	Images for training	Images for testing
V_{s1}	5 % (195)	95 % (3705)	19	1, 2, 3, 4, 5, 6, 7, 8, 9, 10, 11, 12, 13, 14, 15, 16, 17, 18 and 20
V_{s2}	15 % (585)	85 % (3315)	1, 6 and 20	2, 3, 4, 5, 7, 8, 9, 10, 11, 12, 13, 14, 15, 16, 17, 18 and 19
V_{s3}	20 % (780)	80 % (3120)	8, 16, 18 and 19	1, 2, 3, 4, 5, 6, 7, 9, 10, 11, 12, 13, 14, 15, 17 and 20
V_{s4}	70 % (2730)	30 % (1170)	1, 2, 6, 7, 8, 10, 11, 13, 14, 15, 16, 18, 19 and 20	3, 4, 5, 9, 12 and 17
V_{s5}	80 % (3120)	20 % (780)	1, 2, 3, 4, 5, 6, 7, 8, 10, 11, 12, 15, 16, 17, 18 and 20	9, 13, 14 and 19

Table 5.30 External memory of evolution #2 (Essex database)

Granule	Validation					
	V_{s1}	V_{s2}	V_{s3}	V_{s4}	V_{s5}	Average
1	0.2043	0.0424	0.0441	0.0245	0.0147	0.0660
2	0.0939	0.0045	0.0029	0	0	0.0203
3	0.0441	0	0.0010	0	0	0.0090
Average	0.1141	0.0156	0.0160	0.0082	0.0049	0.0318

Table 5.31 Deactivations of evolution #2 (Essex database)

Granule	Validation				
	V_{s1}	V_{s2}	V_{s3}	V_{s4}	V_{s5}
1	–	–	–	–	–
2	–	–	–	6	5
3	–	16	–	5	5

5.2.1   Non-optimized Fuzzy Integration

To combine the responses of the different biometric measures, a non-optimized fuzzy integrator is used to perform a comparison between non-optimized and optimized integration results. The non-optimized fuzzy integrator used is Mamdani type with 81 rules and 3 gaussian membership functions (Low, Medium and High) in each variable; 4 inputs (face, iris, ear and voice) and 1 output (final answer). In Fig. 5.15, the non-optimized fuzzy integrator is shown.

In Table 5.34, the non-optimized fuzzy integration results are shown. In the results can be noticed that the same fuzzy integrator can provide different result in

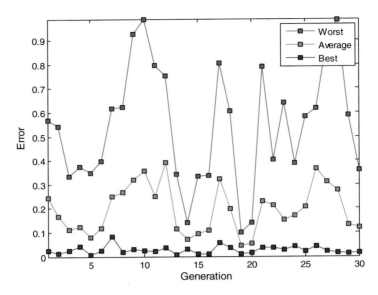

Fig. 5.10 Behavior of evolution # 2 (Essex database)

Table 5.32 Best architectures of evolution #2 (Essex database)

Validation	Granule	Num. Hidden layers and Num. of neurons	Persons per module	Rec. rate (%)	Error
V_{s1}	1	1 (82) 4 (275, 213, 376, 408)	Module #1 (1 to 60) Module #2 (61 to 68)	88.59	0.1141
	2	5 (170, 331, 231, 147, 323) 1 (266) 5 (425, 498, 143, 411, 180) 5 (153, 338, 262, 94, 421) 3 (151,214,261)	Module #3 (69 to 75) Module #4 (76 to 83) Module #5 (84 to 97) Module #6 (98 to 112) Module #7 (113 to 133)		
	3	2 (99, 308) 3 (143, 191, 232) 2 (269, 312) 1 (368)	Module #8 (134 to 149) Module #9 (150 to 165) Module #10 (166 to 173) Module #11 (174 to 195)		
V_{s2}	1	1 (155) 3 (92, 110, 224)	Module #1 (1 to 52) Module #2 (53 to 68)	98.44	0.0156
	2	5 (75, 94, 142, 305, 351) 4 (86, 226, 460, 303) 2 (128, 226) 5 (303, 240, 487, 433, 32)	Module #3 (69 to 88) Module #4 (89 to 95) Module #5 (96 to 118) Module #6 (119 to 133)		
	3	2 (99, 308)	Module #7 (134 to 195)		

(continued)

Table 5.32   (continued)

Validation	Granule	Num. Hidden layers and Num. of neurons	Persons per module	Rec. rate (%)	Error
V_{s3}	1	*1* (90) *2* (46, 250) *5* (307, 428, 268, 220, 201)	Module #1 (1 to 25) Module #2 (26 to 39) Module #3 (40 to 68)	98.40	0.0160
	2	*3* (302, 165, 281) *4* (471, 360, 432, 26) *2* (282, 487)	Module #4 (69 to 84) Module #5 (85 to 110) Module #6 (111 to 133)		
	3	*2* (337, 40) *4* (184, 382, 274, 292) *3* (436, 262, 366) *4* (172, 432, 316, 61)	Module #7 (134 to 156) Module #8 (157 to 166) Module #9 (167 to 177) Module #10 (178 to 195)		
V_{s4}	1	*3* (167, 165, 31) *1* (103)	Module #1 (1 to 34 Module #2 (35 to 68	99.18	0.0082
	2	*2* (122, 380) *1* (303) *2* (203, 275)	Module #3 (69 to 87) Module #4 (88 to 113) Module #5 (114 to 133)		
	3	*5* (325, 328, 23, 214, 212) *4* (464, 461, 341, 177)	Module #6 (134 to 187) Module #7 (188 to 195)		
V_{s5}	1	*3* (24, 234, 478) *4* (403, 329, 51, 153)	Module #1 (1 to 45) Module #2 (46 to 68)	99.51	0.0049
	2	*2* (122, 380) *1* (303) *2* (203, 275)	Module #3 (69 to 102) Module #4 (103 to 111) Module #5 (112 to 133)		
	3	*4* (175, 438, 485, 213) *3* (367, 95, 399)	Module #6 (134 to 163) Module #7 (164 to 195)		

each case. These differences in results can be due to the responses integrated, i.e. each case has different results to be integrated, for this reason a generic fuzzy integrator is not a good option, a better option would be to have an optimal fuzzy integrator depending on the responses that will be integrated.

5.2.2   Optimized Fuzzy Integration

The proposed hierarchical genetic algorithm for fuzzy integrator is used to perform the integration in the cases previously established. To compare the results using the proposed HGA, another optimization is performed. That optimization consists only in; type of fuzzy logic and the parameters of 3 membership functions in each variable (these MFs can be all Gaussian or trapezoidal). We always use 81 fuzzy rules. It is important to remember that the proposed HGA performs the optimization of the type of fuzzy logic, number of MFs in each variable, type de MFs (allows the combination of different types of MFs in the same variable), consequents of the fuzzy rules and the number of fuzzy rules. In Table 5.35, the number of inputs, outputs and the number of possible rules are shown for the proposed HGA.

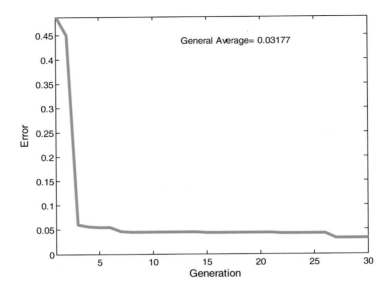

Fig. 5.11 General average of the external memory (evolution #2)

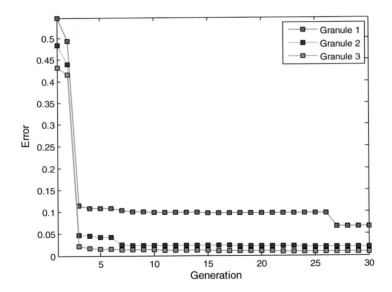

Fig. 5.12 Average of each granule (evolution #2)

In Table 5.36, a comparison among the non-optimized fuzzy integrator shown in Fig. 5.15, the optimization using only 3 membership functions and the optimization performs using the proposed HGA is shown, where the best results can be observed. The results shown, that the proposed method achieved better results than the

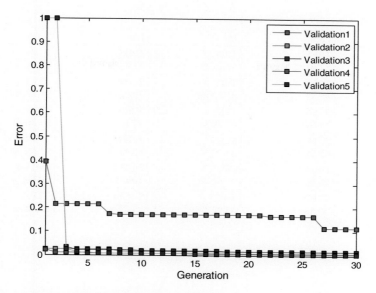

Fig. 5.13 Average of each validation (evolution #2)

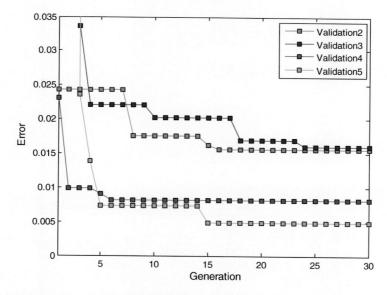

Fig. 5.14 Zoom of the average of each validation (evolution #2)

non-optimized fuzzy integrator and the optimization using a number fixed of membership functions in almost all the cases, only in C5, the same result is obtained in the both optimizations. When the non-optimized fuzzy integrator is used in this case a 99.68 of recognition rate is obtained, for this reason the

Table 5.33 Cases for the fuzzy integration

Case	Face	Iris	Ear	Voice
C1	T2F1 87.01 %	T2I4 82.58 %	T2E2 77.92 %	T2V1 87.79 %
C2	T2F2 85.71 %	T2I2 81.82 %	T2E4 79.22 %	T2V5 91.23 %
C3	T2F6 37.01 %	T2I7 63.20 %	T2E6 57.14 %	T2V10 86.36 %
C4	T2F4 45.78 %	T2I7 63.20 %	T2E9 82.47 %	T2V4 91.88 %
C5	OP1F16 98.05 %	OP1I13 99.68 %	OP1E1 100 %	OP1V4 100 %
C6	T1F26 30.84 %	T4I29 43.92 %	T4E14 77.27 %	T3V22 62.34 %

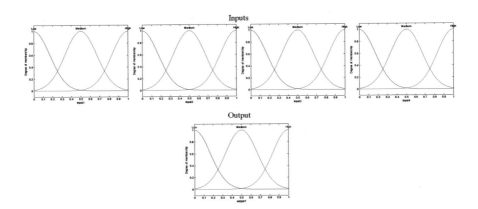

Fig. 5.15 Non-optimized fuzzy integrator

Table 5.34 Non-optimized fuzzy integration results

Case	Face	Iris	Ear	Voice	Non-optimized fuzzy integrator
C1	T2F1 87.01 %	T2I4 82.58 %	T2E2 77.92 %	T2V1 87.79 %	79.22 % 0.2078
C2	T2F2 85.71 %	T2I2 81.82 %	T2E4 79.22 %	T2V5 91.23 %	76.23 % 0.2377
C3	T2F6 37.01 %	T2I7 63.20 %	T2E6 57.14 %	T2V10 86.36 %	58.77 % 0.4123
C4	T2F4 45.78 %	T2I7 63.20 %	T2E9 82.47 %	T2V4 91.88 %	67.53 % 0.3247
C5	OP1F16 98.05 %	OP1I13 99.68 %	OP1E1 100 %	OP1V4 100 %	99.68 % 0.0032
C6	T1F26 30.84 %	T4I29 43.92 %	T4O14 77.27 %	T3V22 62.34 %	43.93 % 0.5207

Table 5.35 Parameters HGA for human recognition (FIS)

Parameter	Value
Number of inputs	4
Number of outputs	1
Maximum number of possible rules	625

Table 5.36 Comparison of results (FUZZY integration)

Case	Non-optimized fuzzy integrator	3 MFs (optimized)	Proposed method
C1	79.22 % 0.2078	92.53 % 0.0743	95.67 % 0.0433
C2	76.23 % 0.2377	93.90 % 0.0610	98.96 % 0.0104
C3	58.77 % 0.4123	88.96 % 0.1104	90.69 % 0.0931
C4	67.53 % 0.3247	95.89 % 0.0411	96.86 % 0.0314
C5	99.68 % 0.0032	100 % 0	100 % 0
C6	43.93 % 0.5207	79.81 % 0.2019	82.88 % 0.1712

improvement is not a challenge for any of both fuzzy inferences system optimizations used. The best evolutions of each case are presented below.

In Figs. 5.16, 5.17 and 5.18, the convergence, the fuzzy integrator and the fuzzy rules of the best evolution for the case C1 are respectively shown. The Fuzzy integrator uses 19 of 24 fuzzy rules and is of Sugeno type.

In Figs. 5.19 and 5.20, the convergence and the fuzzy integrator of the best evolution for the case C2 are respectively shown. The Fuzzy integrator uses 108 of 144 fuzzy rules and is of Sugeno type.

In Figs. 5.21, 5.22 and 5.23, the convergence and the fuzzy integrator and the fuzzy rules of the best evolution for the case C3 are respectively shown. The Fuzzy integrator uses 13 of 24 fuzzy rules and is of Sugeno type.

In Figs. 5.24 and 5.25, the convergence and the fuzzy integrator of the best evolution for the case C4 are respectively shown. The Fuzzy integrator uses 44 of 96 fuzzy rules and is of Sugeno type.

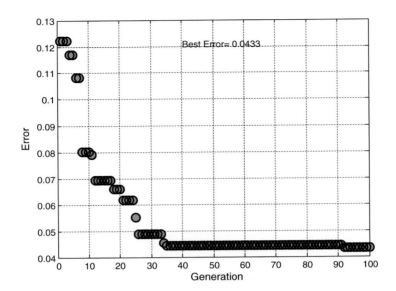

Fig. 5.16 Convergence of the best prediction for the case C1

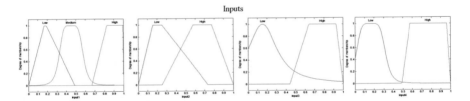

Fig. 5.17 Best fuzzy integrator for the case C1

1. If (input1 is Low) and (input2 is Low) and (input3 is Low) and (input4 is Low) then (output1 is MediumLow) (1)
2. If (input1 is Low) and (input2 is Low) and (input3 is High) and (input4 is Low) then (output1 is Low) (1)
3. If (input1 is Low) and (input2 is Low) and (input3 is High) and (input4 is High) then (output1 is MediumHigh) (1)
4. If (input1 is Low) and (input2 is High) and (input3 is Low) and (input4 is Low) then (output1 is Low) (1)
5. If (input1 is Low) and (input2 is High) and (input3 is Low) and (input4 is High) then (output1 is MediumHigh) (1)
6. If (input1 is Low) and (input2 is High) and (input3 is High) and (input4 is High) then (output1 is High) (1)
7. If (input1 is Medium) and (input2 is Low) and (input3 is Low) and (input4 is High) then (output1 is Low) (1)
8. If (input1 is Medium) and (input2 is Low) and (input3 is High) and (input4 is Low) then (output1 is High) (1)
9. If (input1 is Medium) and (input2 is Low) and (input3 is High) and (input4 is High) then (output1 is MediumHigh) (1)
10. If (input1 is Medium) and (input2 is High) and (input3 is Low) and (input4 is Low) then (output1 is MediumHigh) (1)
11. If (input1 is Medium) and (input2 is High) and (input3 is Low) and (input4 is High) then (output1 is Low) (1)
12. If (input1 is Medium) and (input2 is High) and (input3 is High) and (input4 is Low) then (output1 is High) (1)
13. If (input1 is High) and (input2 is Low) and (input3 is Low) and (input4 is Low) then (output1 is Low) (1)
14. If (input1 is High) and (input2 is Low) and (input3 is Low) and (input4 is High) then (output1 is High) (1)
15. If (input1 is High) and (input2 is Low) and (input3 is High) and (input4 is Low) then (output1 is Low) (1)
16. If (input1 is High) and (input2 is Low) and (input3 is High) and (input4 is High) then (output1 is High) (1)
17. If (input1 is High) and (input2 is High) and (input3 is Low) and (input4 is Low) then (output1 is MediumLow) (1)
18. If (input1 is High) and (input2 is High) and (input3 is Low) and (input4 is High) then (output1 is High) (1)
19. If (input1 is High) and (input2 is High) and (input3 is High) and (input4 is Low) then (output1 is MediumLow) (1)

Fig. 5.18 Fuzzy rules of the best fuzzy integrator for the case C1

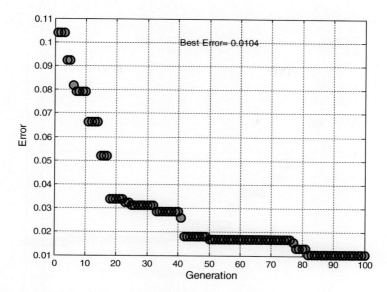

Fig. 5.19 Convergence of the best prediction for the case C2

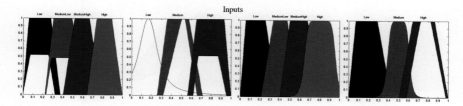

Fig. 5.20 Best fuzzy integrator for the case C2

In Figs. 5.26 and 5.27, the convergence and the fuzzy integrator of the best evolution for the case C5 are respectively shown. This fuzzy integrator was obtained with the optimization performed of 3 MFs, for this reason this fuzzy integrator uses all the possible fuzzy rules (81 fuzzy rules) and is of Sugeno Type.

In Figs. 5.28 and 5.29, the convergence and the fuzzy integrator of the best evolution for the case C6 are respectively shown. The Fuzzy integrator uses 49 of 90 fuzzy rules and is of Sugeno type.

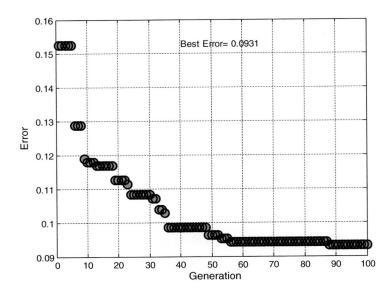

Fig. 5.21 Convergence of the best prediction for the case C3

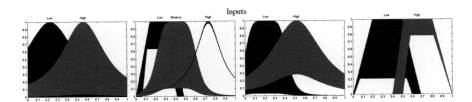

Fig. 5.22 Best fuzzy integrator for the case C3

```
1. If (input1 is Low) and (input2 is Low) and (input3 is Low) and (input4 is Low) then (output1 is High) (1)
2. If (input1 is Low) and (input2 is Low) and (input3 is High) and (input4 is Low) then (output1 is MediumHigh) (1)
3. If (input1 is Low) and (input2 is Low) and (input3 is High) and (input4 is High) then (output1 is MediumHigh) (1)
4. If (input1 is Low) and (input2 is Medium) and (input3 is High) and (input4 is High) then (output1 is MediumLow) (1)
5. If (input1 is Low) and (input2 is High) and (input3 is Low) and (input4 is Low) then (output1 is MediumLow) (1)
6. If (input1 is Low) and (input2 is High) and (input3 is Low) and (input4 is High) then (output1 is High) (1)
7. If (input1 is Low) and (input2 is High) and (input3 is High) and (input4 is Low) then (output1 is Low) (1)
8. If (input1 is Low) and (input2 is High) and (input3 is High) and (input4 is High) then (output1 is High) (1)
9. If (input1 is High) and (input2 is Low) and (input3 is Low) and (input4 is Low) then (output1 is High) (1)
10. If (input1 is High) and (input2 is Low) and (input3 is High) and (input4 is Low) then (output1 is MediumLow) (1)
11. If (input1 is High) and (input2 is Medium) and (input3 is Low) and (input4 is Low) then (output1 is MediumLow) (1)
12. If (input1 is High) and (input2 is High) and (input3 is Low) and (input4 is Low) then (output1 is MediumHigh) (1)
13. If (input1 is High) and (input2 is High) and (input3 is Low) and (input4 is High) then (output1 is MediumHigh) (1)
```

Fig. 5.23 Fuzzy rules of the best fuzzy integrator for the case C3

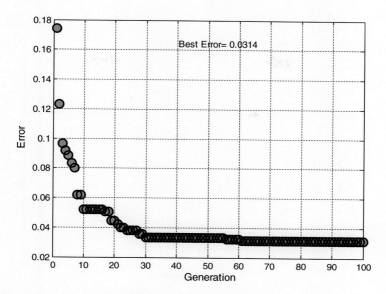

Fig. 5.24 Convergence of the best prediction for the case C4

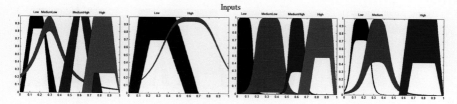

Fig. 5.25 Best fuzzy integrator for the case C4

5.3   Summary of Results

In this section, a summary of results obtained with the different tests and optimization are presented below.

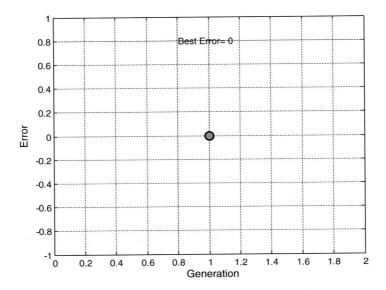

Fig. 5.26 Convergence of the best prediction for the case C5

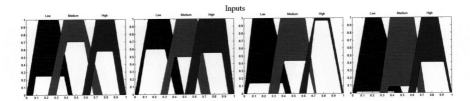

Fig. 5.27 Best fuzzy integrator for the case C5

5.3.1   Comparisons of Granulation #1

In Table 5.37, a summary of average of results with the different test and optimizations using the first granulation is shown.

5.3.2   Comparisons of Granulation #2

In Table 5.38, a summary of average of results of the Essex Database using the
second granulation is shown.

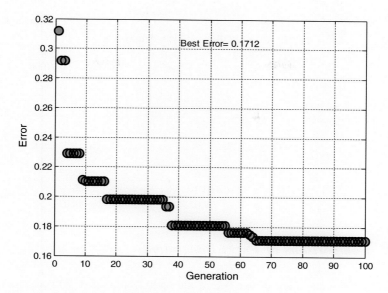

Fig. 5.28 Convergence of the best prediction for the case C6

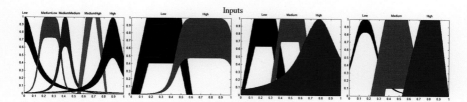

Fig. 5.29 Best fuzzy integrator for the case C6

Table 5.37 Average of results (biometric measures)

Biometric measure	T1	T2	T3	T4	OP1	OP2
Face	55.71 % 0.4429	67.77 % 0.3223	50.36 % 0.4964	66.54 % 0.3346	96.09 % 0.0391	93.13 % 0.0687
Iris	85.84 % 0.1416	86.40 % 0.1360	72.20 % 0.2780	76.00 % 0.2400	98.68 % 0.0132	96.48 % 0.0352
Ear	66.62 % 0.3338	72.56 % 0.2744	59.13 % 0.4087	59.52 % 0.4048	99.61 % 0.0039	94.40 % 0.0560
Voice	82.57 % 0.1743	86.98 % 0.1302	78.59 % 0.2141	83.68 % 0.1632	99.44 % 0.0056	97.43 % 0.0257

Table 5.38 Table of
comparison for Essex
database results

Non-optimized		Granulation #2
Color RGB	Type-1 fuzzy logic	
46.82 % 0.5318	53.21 % 0.4679	98.09 % 0.01906

Table 5.39 Comparison of
results (fuzzy integration)

Case	3 MFs (optimized) (%)	Proposed method (%)
C1	87.83	94.20
C2	90.99	97.67
C3	84.42	88.94
C4	92.67	95.83
C5	100	100
C6	77.02	80.07

5.3.3 Comparisons of Fuzzy Integration

In Table 5.39, a summary of average of the optimization of 3 MFs and the optimization performs using the proposed HGA is shown.

Chapter 6
Conclusions

In this book, a new method of modular granular neural networks and their integrations using fuzzy logic is proposed. The main contribution is the development of a method based on granular computing where, the granulation is used in 2 levels. The information or data used in the modular neural networks is granulated to improve the effectiveness of these MNNs, and the integration of responses is performed using a fuzzy inference system where, the granulation is also performed. Different tests (non-optimized) were performed, where the number of modules and data for the training phase were modified. Optimizations were also performed to improve the results obtained and better results were achieved. The combination of responses is performed using fuzzy logic where a hierarchical genetic algorithm was developed to find the optimal fuzzy integrator parameters. Better results were obtained using the optimizations developed in this research work. A new method to work with big databases was also proposed, where based on the database complexity a optimization is performed.

As future works, the granulation performed using the database complexity in this research work was only used with a biometric measure. The future work for this part would be, if more biometric measures are used, how to integrate each biometric measure of a persons, if for example, in the case of face, if a person has a medium complexity level, in the iris has a high complexity level, in the ear a low complexity level and with voice has a high complexity level. The question would be, what is his final new ID? An answer would perhaps be to find a method (maybe a fuzzy inference system), where the inputs would be the percentage of complexity of each biometric measure, logically the number of inputs would be the number of biometric measured used, and a final percentage of complexity is obtained, and so, a final new ID is obtained for each person and the optimization based on the database complexity is performed for each biometric measure.

The proposed HGA for fuzzy inference system optimization can be easily applied to another application, where a fuzzy inference system is used. Finally, the proposed granulations in this book can be adapted to other optimization methods such as; particle swarm optimization, ant colony system or another bio inspired optimization algorithm.

© The Author(s) 2016
D. Sanchez and P. Melin, *Hierarchical Modular Granular Neural Networks with Fuzzy Aggregation*, SpringerBriefs in Computational Intelligence,
DOI 10.1007/978-3-319-28862-8_6

Appendix A
Appendix of Biometric Measures

The results obtained for each biometric measure in the optimizations are presented.

A.1 Face (CASIA Database)

The results obtained for face (CASIA Database) in the Test #1, Test #2, Test #3, Test #4, the optimization using up to 80 and 50 % of data for trainings are respectively presented in Tables A.1, A.2, A.3, A.4, A.5 and A.6.

A.2 Iris

The results obtained for iris in the Test #1, Test #2, Test #3, Test #4, the optimization using up to 80 and 50 % of data for trainings are respectively presented in Tables A.7, A.8, A.9, A.10, A.11 and A.12.

A.3 Ear

The results obtained for ear in the Test #1, Test #2, Test #3, Test #4, the optimization using up to 80 and 50 % of data for trainings are respectively presented in Tables A.13, A.14, A.15, A.16, A.17 and A.18.

A.4 Voice

The results obtained for voice in the Test #1, Test #2, Test #3, Test #4, the optimization using up to 80 and 50 % of data for trainings are respectively presented in Tables A.19, A.20, A.21, A.22, A.23 and A.24.

© The Author(s) 2016
D. Sanchez and P. Melin, *Hierarchical Modular Granular Neural Networks with Fuzzy Aggregation*, SpringerBriefs in Computational Intelligence, DOI 10.1007/978-3-319-28862-8

Table A.1 Non-optimized trainings (with 3 modules, Face, 80 %)

Training	Images for training (%)	Rec. Rate (%)	Error Rec.	Training	Images for training (%)	Rec. Rate (%)	Error Rec.
T1F1	38	61.04	0.3896	T1F16	23	53.25	0.4675
T1F2	35	60.61	0.3939	T1F17	18	27.60	0.7240
T1F3	78	71.43	0.2857	T1F18	48	71.43	0.2857
T1F4	72	89.61	0.1039	T1F19	2	44.81	0.5519
T1F5	43	70.13	0.2987	T1F20	33	57.14	0.4286
T1F6	26	63.31	0.3669	T1F21	53	53.25	0.4675
T1F7	48	63.20	0.3680	T1F22	13	47.40	0.5260
T1F8	58	57.79	0.4221	T1F23	27	44.16	0.5584
T1F9	33	70.13	0.2987	T1F24	36	72.29	0.2771
T1F10	12	38.96	0.6104	T1F25	81	77.92	0.2208
T1F11	23	50.97	0.4903	T1F26	16	30.84	0.6916
T1F12	9	25.32	0.7468	T1F27	8	46.43	0.5357
T1F13	42	63.64	0.3636	T1F28	48	64.94	0.3506
T1F14	8	51.95	0.4805	T1F29	3	29.87	0.7013
T1F15	4	49.03	0.5097	T1F30	30	62.77	0.3723

Table A.2 Non-optimized trainings (with different number of modules, Face, 80 %)

Training	Modules	Images for training (%)	Rec. Rate (%)	Error Rec.	Training	Modules	Images for training (%)	Rec. Rate (%)	Error Rec.
T2F1	6	79	87.01	0.1299	T2F16	5	60	72.08	0.2792
T2F2	10	69	85.71	0.1429	T2F17	6	44	63.64	0.3636
T2F3	9	19	52.92	0.4708	T2F18	8	20	67.86	0.3214
T2F4	6	11	45.78	0.5422	T2F19	6	9	51.30	0.4870
T2F5	7	49	60.17	0.3983	T2F20	6	18	67.86	0.3214
T2F6	5	13	37.01	0.6299	T2F21	3	68	80.52	0.1948
T2F7	4	30	47.19	0.5281	T2F22	3	22	55.52	0.4448
T2F8	6	50	70.78	0.2922	T2F23	2	68	80.52	0.1948
T2F9	5	77	68.83	0.3117	T2F24	8	38	73.16	0.2684
T2F10	2	50	68.83	0.3117	T2F25	9	32	62.34	0.3766
T2F11	7	38	72.29	0.2771	T2F26	7	77	64.94	0.3506
T2F12	3	54	71.43	0.2857	T2F27	8	48	72.73	0.2727
T2F13	6	36	61.47	0.3853	T2F28	2	54	88.31	0.1169
T2F14	4	51	88.31	0.1169	T2F29	3	32	69.26	0.3074
T2F15	8	36	75.76	0.2424	T2F30	4	56	69.48	0.3052

Table A.3 Non-optimized trainings (with 3 modules, Face, 50 %)

Training	Images for training (%)	Rec. Rate (%)	Error Rec.	Training	Images for training (%)	Rec. Rate (%)	Error Rec.
T3F1	47	58.01	0.4199	T3F16	9	37.99	0.6201
T3F2	6	57.14	0.4286	T3F17	23	36.04	0.6396
T3F3	18	45.13	0.5487	T3F18	10	41.56	0.5844
T3F4	30	42.86	0.5714	T3F19	23	28.57	0.7143
T3F5	16	50.97	0.4903	T3F20	43	73.59	0.2641
T3F6	4	39.61	0.6039	T3F21	48	69.26	0.3074
T3F7	46	64.07	0.3593	T3F22	25	51.30	0.4870
T3F8	43	68.40	0.3160	T3F23	38	61.04	0.3896
T3F9	47	41.99	0.5801	T3F24	8	39.29	0.6071
T3F10	35	63.64	0.3636	T3F25	26	43.18	0.5682
T3F11	38	66.67	0.3333	T3F26	34	67.10	0.3290
T3F12	35	52.38	0.4762	T3F27	16	46.10	0.5390
T3F13	15	35.06	0.6494	T3F28	20	44.16	0.5584
T3F14	27	55.84	0.4416	T3F29	3	37.66	0.6234
T3F15	7	48.70	0.5130	T3F30	3	43.51	0.5649

Table A.4 Non-optimized trainings (with different number of modules, Face, 50 %)

Training	Modules	Images for training (%)	Rec. Rate (%)	Error Rec.	Training	Modules	Images for training (%)	Rec. Rate (%)	Error Rec.
T4F1	9	38	76.19	0.2381	T4F16	8	6	54.87	0.4513
T4F2	7	34	77.49	0.2251	T4F17	9	21	71.43	0.2857
T4F3	3	44	63.20	0.3680	T4F18	3	51	89.61	0.1039
T4F4	5	36	64.94	0.3506	T4F19	10	26	69.81	0.3019
T4F5	7	3	39.94	0.6006	T4F20	9	20	67.53	0.3247
T4F6	3	41	55.41	0.4459	T4F21	9	34	74.46	0.2554
T4F7	6	37	67.10	0.3290	T4F22	10	39	76.62	0.2338
T4F8	6	10	64.94	0.3506	T4F23	10	50	91.56	0.0844
T4F9	7	29	65.26	0.3474	T4F24	9	40	83.98	0.1602
T4F10	4	24	56.49	0.4351	T4F25	2	33	49.78	0.5022
T4F11	4	37	67.53	0.3247	T4F26	10	9	59.42	0.4058
T4F12	5	2	63.31	0.3669	T4F27	8	41	77.92	0.2208
T4F13	9	35	67.10	0.3290	T4F28	9	37	71.86	0.2814
T4F4	8	30	76.19	0.2381	T4F29	2	7	49.68	0.5032
T4F15	2	6	27.27	0.7273	T4F30	8	37	75.32	0.2468

Table A.5 Optimized results (Face, 80 %)

Evolution	Modules	Images for training (%)	Rec. Rate (%)	Error Rec.	Evolution	Modules	Images for training (%)	Rec. Rate (%)	Error Rec.
OP1F1	10	50	96.75	0.0325	OP1F11	8	50	96.75	0.0325
OP1F2	2	50	95.45	0.0455	OP1F12	4	50	96.10	0.0390
OP1F3	6	50	96.10	0.0390	OP1F13	2	50	96.10	0.0390
OP1F4	10	50	97.40	0.0260	OP1F14	2	50	95.45	0.0455
OP1F5	6	50	96.10	0.0390	OP1F15	4	50	95.45	0.0455
OP1F6	5	50	96.75	0.0325	OP1F16	6	50	98.05	0.0195
OP1F7	7	50	96.75	0.0325	OP1F17	10	50	97.40	0.0260
OP1F8	10	50	95.45	0.0455	OP1F18	8	30	87.45	0.1255
OP1F9	8	50	97.40	0.0260	OP1F19	6	50	96.75	0.0325
OP1F10	8	50	96.75	0.0325	OP1F20	10	50	97.40	0.0260

Table A.6 Optimized results (Face, 50 %)

Evolution	Modules	Images for training (%)	Rec. Rate (%)	Error Rec.	Evolution	Modules	Images for training (%)	Rec. Rate (%)	Error Rec.
OP2F1	10	50	96.75	0.0325	OP2F11	8	50	96.75	0.0325
OP2F2	2	50	95.45	0.0455	OP2F12	4	50	96.10	0.0390
OP2F3	6	50	96.10	0.0390	OP2F13	2	50	96.10	0.0390
OP2F4	10	50	97.40	0.0260	OP2F14	6	41	87.01	0.1299
OP2F5	6	50	96.10	0.0390	OP2F15	5	31	86.58	0.1342
OP2F6	5	50	96.75	0.0325	OP2F16	8	31	88.31	0.1169
OP2F7	7	50	96.75	0.0325	OP2F17	8	43	86.58	0.1342
OP2F8	10	50	95.45	0.0455	OP2F18	3	40	83.55	0.1645
OP2F9	8	50	97.40	0.0260	OP2F19	9	34	88.31	0.1169
OP2F10	8	50	96.75	0.0325	OP2F20	8	49	88.31	0.1169

A.5   Face (Essex Database)

The results of the non-optimized trainings are shown in Tables A.25 and A.26. The different external memories achieved in the 5 evolutions using the method based on the database complexity are respectively presented in Tables A.27, A.28, A.29, A.30 and A.31.

Table A.7 Non-optimized trainings (with 3 modules, Iris, 80 %)

Training	Images for training (%)	Rec. Rate (%)	Error Rec.	Training	Images for training (%)	Rec. Rate (%)	Error Rec.
T1I1	67	94.03	0.0597	T1I16	64	92.73	0.0727
T1I2	20	80.76	0.1924	T1I17	39	80.95	0.1905
T1I3	53	90.72	0.0928	T1I18	2	40.86	0.5914
T1I4	50	90.54	0.0946	T1I19	52	85.16	0.1484
T1I5	43	91.07	0.0893	T1I20	24	78.98	0.2102
T1I6	80	94.81	0.0519	T1I21	56	93.51	0.0649
T1I7	31	84.68	0.1532	T1I22	40	93.18	0.0682
T1I8	43	86.69	0.1331	T1I23	53	92.39	0.0761
T1I9	60	92.21	0.0779	T1I24	68	94.48	0.0552
T1I10	42	88.47	0.1153	T1I25	64	94.55	0.0545
T1I11	66	92.99	0.0701	T1I26	69	83.77	0.1623
T1I12	47	91.47	0.0853	T1I27	32	78.44	0.2156
T1I13	11	55.52	0.4448	T1I28	5	54.65	0.4535
T1I14	58	93.51	0.0649	T1I29	77	95.24	0.0476
T1I15	67	92.73	0.0727	T1I30	74	96.10	0.0390

Table A.8 Optimized Trainings (with different number of modules, Iris, 80 %)

Training	Modules	Images for training (%)	Rec. Rate (%)	Error Rec.	Training	Modules	Images for training (%)	Rec. Rate (%)	Error Rec.
T2I1	4	19	79.10	0.2090	T2I16	2	34	88.31	0.1169
T2I2	3	27	81.82	0.1818	T2I17	3	58	93.94	0.0606
T2I3	2	73	96.10	0.0390	T2I18	2	30	80.91	0.1909
T2I4	10	11	82.58	0.1742	T2I19	5	19	82.64	0.1736
T2I5	8	56	94.37	0.0563	T2I20	5	59	91.13	0.0887
T2I6	1	71	90.91	0.0909	T2I21	2	25	72.21	0.2779
T2I7	4	16	63.20	0.3680	T2I22	2	16	59.85	0.4015
T2I8	9	16	84.96	0.1504	T2I23	6	76	96.97	0.0303
T2I9	5	61	92.73	0.0727	T2I24	9	59	94.37	0.0563
T2I10	7	80	98.27	0.0173	T2I25	4	38	87.45	0.1255
T2I11	6	22	63.40	0.3660	T2I26	2	75	93.51	0.0649
T2I12	8	75	96.97	0.0303	T2I27	5	18	83.23	0.1677
T2I13	4	72	93.83	0.0617	T2I28	3	16	76.73	0.2327
T2I14	10	43	94.48	0.0552	T2I29	7	74	95.78	0.0422
T2I15	10	66	97.66	0.0234	T2I30	5	31	84.55	0.1545

Table A.9 Non-optimized trainings (with 3 modules, Iris, 50 %)

Training	Images for training (%)	Rec. Rate (%)	Error Rec.	Training	Images for training (%)	Rec. Rate (%)	Error Rec.
T3I1	25	81.69	0.1831	T3I16	5	49.85	0.5015
T3I2	21	80.64	0.1936	T3I17	12	50.97	0.4903
T3I3	46	89.29	0.1071	T3I18	44	81.33	0.1867
T3I4	11	74.13	0.2587	T3I19	28	88.70	0.1130
T3I5	42	89.45	0.1055	T3I20	18	71.43	0.2857
T3I6	18	76.51	0.2349	T3I21	29	76.36	0.2364
T3I7	32	77.01	0.2299	T3I22	41	89.45	0.1055
T3I8	40	87.34	0.1266	T3I23	21	48.17	0.5183
T3I9	24	74.14	0.2586	T3I24	14	72.40	0.2760
T3I10	30	82.73	0.1727	T3I25	19	83	0.1700
T3I11	12	75.32	0.2468	T3I26	5	40.96	0.5904
T3I12	7	52.75	0.4725	T3I27	33	89.03	0.1097
T3I13	24	75.56	0.2444	T3I28	8	47.85	0.5215
T3I14	29	81.17	0.1883	T3I29	25	79.74	0.2026
T3I15	8	47.05	0.5295	T3I30	2	51.85	0.4815

Table A.10 Non-optimized trainings (with different number of modules, Iris, 50 %)

Training	Modules	Images for training (%)	Rec. Rate (%)	Error Rec.	Training	Modules	Images for training (%)	Rec. Rate (%)	Error Rec.
T4I1	10	10	65.53	0.3447	T4I16	6	29	88.18	0.1182
T4I2	4	4	51.75	0.4825	T4I17	9	47	94.25	0.0575
T4I3	9	41	93.18	0.0682	T4I18	8	46	84.09	0.1591
T4I4	5	8	53.65	0.4635	T4I19	9	51	94.25	0.0575
T4I5	2	51	83.30	0.1670	T4I20	9	48	95.55	0.0445
T4I6	8	45	90.75	0.0925	T4I21	2	32	79.22	0.2078
T4I7	5	36	90.33	0.0967	T4I22	8	12	78.46	0.2154
T4I8	3	19	69.07	0.3093	T4I23	5	4	53.85	0.4615
T4I9	10	3	71.33	0.2867	T4I24	9	17	70.67	0.2933
T4I10	4	22	75.09	0.2491	T4I25	7	39	87.30	0.1270
T4I11	5	25	84.94	0.1506	T4I26	2	41	73.70	0.2630
T4I12	9	39	86.15	0.1385	T4I27	5	21	83.71	0.1629
T4I13	8	29	70	0.3000	T4I28	9	7	55.84	0.4416
T4I14	10	14	82.79	0.1721	T4I29	2	18	43.92	0.5608
T4I15	3	35	84.70	0.1530	T4I30	2	5	44.46	0.5554

Table A.11 Optimized results (Iris, 80 %)

Evolution	Modules	Images for training (%)	Rec. Rate (%)	Error Rec.	Evolution	Modules	Images for training (%)	Rec. Rate (%)	Error Rec.
OP1I1	2	78	97.84	0.0216	OP1I11	6	79	99.57	0.0043
OP1I2	9	72	98.70	0.0130	OP1I12	8	63	98.44	0.0156
OP1I3	7	66	97.92	0.0208	OP1I13	9	71	99.68	0.0032
OP1I4	9	69	99.35	0.0065	OP1I14	7	71	99.35	0.0065
OP1I5	2	78	97.84	0.0216	OP1I15	3	61	97.40	0.0260
OP1I6	9	72	98.70	0.0130	OP1I16	9	79	99.57	0.0043
OP1I7	7	66	97.92	0.0208	OP1I17	6	55	98.05	0.0195
OP1I8	9	69	99.35	0.0065	OP1I18	8	69	99.35	0.0065
OP1I9	10	63	98.96	0.0104	OP1I19	6	75	99.57	0.0043
OP1I10	8	41	97.40	0.0260	OP1I20	7	68	98.70	0.0130

Table A.12 Optimized results (Iris, 50 %)

Evolution	Modules	Images for training (%)	Rec. Rate (%)	Error Rec.	Evolution	Modules	Images for training (%)	Rec. Rate (%)	Error Rec.
OP2I1	8	50	97.03	0.0297	OP2I11	5	41	96.10	0.0390
OP2I2	10	39	96.10	0.0390	OP2I12	8	50	97.03	0.0297
OP2I3	4	42	94.64	0.0536	OP2I13	8	37	94.95	0.0505
OP2I4	8	48	97.40	0.0260	OP2I14	7	49	96.85	0.0315
OP2I5	9	50	97.40	0.0260	OP2I15	7	36	95.82	0.0418
OP2I6	6	45	96.75	0.0325	OP2I16	7	47	97.40	0.0260
OP2I7	8	39	96.10	0.0390	OP2I17	9	48	97.03	0.0297
OP2I8	6	47	96.85	0.0315	OP2I18	9	41	96.27	0.0373
OP2I9	7	39	94.66	0.0534	OP2I19	10	48	97.77	0.0223
OP2I10	10	46	97.24	0.0276	OP2I20	6	42	96.27	0.0373

Table A.13 Non-optimized Trainings (with 3 modules, Ear, 80 %)

Training	Images for training (%)	Rec. Rate (%)	Error Rec.	Training	Images for training (%)	Rec. Rate (%)	Error Rec.
T1E1	68	53.25	0.4675	T1E16	52	84.42	0.1558
T1E2	70	97.40	0.0260	T1E17	70	75.32	0.2468
T1E3	18	57.58	0.4242	T1E18	34	66.67	0.3333
T1E4	17	61.47	0.3853	T1E19	13	43.29	0.5671
T1E5	59	66.23	0.3377	T1E20	76	97.40	0.0260
T1E6	2	43.72	0.5628	T1E21	37	50.65	0.4935
T1E7	47	71.43	0.2857	T1E22	69	62.34	0.3766
T1E8	34	35.50	0.6450	T1E23	42	81.82	0.1818
T1E9	58	74.03	0.2597	T1E24	28	46.32	0.5368
T1E10	52	72.73	0.2727	T1E25	49	78.57	0.2143
T1E11	32	49.78	0.5022	T1E26	74	98.70	0.0130
T1E12	72	61.04	0.3896	T1E27	64	72.73	0.2727
T1E13	58	83.12	0.1688	T1E28	22	45.45	0.5455
T1E14	52	69.48	0.3052	T1E29	61	70.13	0.2987
T1E15	49	75.97	0.2403	T1E30	24	51.95	0.4805

Table A.14 Non-optimized trainings (with different number of modules, Ear, 80 %)

Training	Modules	Images for training (%)	Rec. Rate (%)	Error Rec.	Training	Modules	Images for training (%)	Rec. Rate (%)	Error Rec.
T2E1	9	74	94.81	0.0519	T2E16	10	59	88.31	0.1169
T2E2	5	66	77.92	0.2208	T2E17	7	16	66.23	0.3377
T2E3	9	79	96.10	0.0390	T2E18	5	54	83.12	0.1688
T2E4	9	51	79.22	0.2078	T2E19	4	10	48.48	0.5152
T2E5	6	81	97.40	0.0260	T2E20	5	55	74.68	0.2532
T2E6	8	27	57.14	0.4286	T2E21	1	62	61.69	0.3831
T2E7	5	57	81.82	0.1818	T2E22	5	69	96.10	0.0390
T2E8	3	66	90.91	0.0909	T2E23	3	29	39.83	0.6017
T2E9	4	56	82.47	0.1753	T2E24	3	65	63.64	0.3636
T2E10	4	38	67.53	0.3247	T2E25	5	48	73.38	0.2662
T2E11	9	24	60.61	0.3939	T2E26	9	14	70.13	0.2987
T2E12	3	9	42.86	0.5714	T2E27	2	54	53.90	0.4610
T2E13	7	50	75.32	0.2468	T2E28	3	77	92.21	0.0779
T2E14	7	44	72.73	0.2727	T2E29	9	70	59.74	0.4026
T2E15	7	47	71.43	0.2857	T2E30	2	21	57.14	0.4286

Table A.15 Non-optimized trainings (with 3 modules, Ear, 50 %)

Training	Images for training (%)	Rec. Rate (%)	Error Rec.	Training	Images for training (%)	Rec. Rate (%)	Error Rec.
T3E1	30	66.23	0.3377	T3E16	17	65.80	0.3420
T3E2	46	65.58	0.3442	T3E17	3	50.65	0.4935
T3E3	30	46.32	0.5368	T3E18	26	43.29	0.5671
T3E4	38	66.23	0.3377	T3E19	48	65.58	0.3442
T3E5	35	59.31	0.4069	T3E20	12	50.65	0.4935
T3E6	20	34.20	0.6580	T3E21	4	52.81	0.4719
T3E7	45	77.92	0.2208	T3E22	33	43.29	0.5671
T3E8	31	59.74	0.4026	T3E23	16	54.55	0.4545
T3E9	45	85.06	0.1494	T3E24	7	51.08	0.4892
T3E10	49	68.83	0.3117	T3E25	32	51.52	0.4848
T3E11	39	70.13	0.2987	T3E26	24	52.38	0.4762
T3E12	48	66.23	0.3377	T3E27	32	64.50	0.3550
T3E13	48	70.78	0.2922	T3E28	34	53.25	0.4675
T3E14	11	39.83	0.6017	T3E29	42	87.66	0.1234
T3E15	24	50.65	0.4935	T3E30	22	59.74	0.4026

Table A.16 Non-optimized trainings (with different number of modules, Ear, 50 %)

Training	Modules	Images for training (%)	Rec. Rate (%)	Error Rec.	Training	Modules	Images for training (%)	Rec. Rate (%)	Error Rec.
T4E1	4	8	56.28	0.4372	T4E16	9	4	55.84	0.4416
T4E2	4	3	59.74	0.4026	T4E17	10	34	57.58	0.4242
T4E3	8	33	54.55	0.4545	T4E18	2	30	51.08	0.4892
T4E4	2	36	44.59	0.5541	T4E19	2	4	56.28	0.4372
T4E5	6	30	54.11	0.4589	T4E20	5	44	64.94	0.3506
T4E6	2	4	41.99	0.5801	T4E21	10	14	53.68	0.4632
T4E7	7	17	64.94	0.3506	T4E22	7	13	54.11	0.4589
T4E8	6	45	79.87	0.2013	T4E23	5	48	78.57	0.2143
T4E9	10	23	58.44	0.4156	T4E24	6	14	56.28	0.4372
T4E10	2	50	66.23	0.3377	T4E25	4	8	49.35	0.5065
T4E11	2	32	42.42	0.5758	T4E26	7	36	73.59	0.2641
T4E12	8	35	54.98	0.4502	T4E27	9	9	63.64	0.3636
T4E13	7	22	62.77	0.3723	T4E28	2	19	42.42	0.5758
T4E14	7	47	77.27	0.2273	T4E29	4	48	56.49	0.4351
T4E15	8	2	70.56	0.2944	T4E30	4	47	83.12	0.1688

Table A.17 Optimized results (Ear, 80 %)

Evolution	Modules	Images for training (%)	Rec. Rate (%)	Error Rec.	Evolution	Modules	Images for training (%)	Rec. Rate (%)	Error Rec.
OP1E1	8	79	100	0	OP1E11	6	75	100	0
OP1E2	8	66	100	0	OP1E12	4	66	100	0
OP1E3	9	72	100	0	OP1E13	10	77	100	0
OP1E4	4	70	100	0	OP1E14	9	71	100	0
OP1E5	2	78	98.70	0.0130	OP1E15	6	66	100	0
OP1E6	2	80	100	0	OP1E16	6	80	100	0
OP1E7	10	70	100	0	OP1E17	8	72	100	0
OP1E8	7	75	100	0	OP1E18	3	59	93.50	0.0649
OP1E9	8	67	100	0	OP1E19	6	75	100	0
OP1E10	9	71	100	0	OP1E20	8	71	100	0

Table A.18 Optimized results (Ear, 50 %)

Evolution	Modules	Images for training (%)	Rec. Rate (%)	Error Rec.	Evolution	Modules	Images for training (%)	Rec. Rate (%)	Error Rec.
OP2E1	6	41	95.45	0.0455	OP2E11	6	41	95.45	0.0455
OP2E2	6	48	94.15	0.0584	OP2E12	6	47	95.45	0.0455
OP2E3	6	41	97.4	0.0260	OP2E13	6	41	96.10	0.0390
OP2E4	6	50	96.10	0.0390	OP2E14	6	39	96.10	0.0390
OP2E5	10	49	96.10	0.0390	OP2E15	6	41	94.80	0.0519
OP2E6	6	41	95.45	0.0455	OP2E16	8	48	96.75	0.0325
OP2E7	8	14	79.65	0.2035	OP2E17	6	41	95.45	0.0455
OP2E8	6	41	96.10	0.0390	OP2E18	8	47	95.45	0.0455
OP2E9	10	45	95.45	0.0455	OP2E19	10	50	86.36	0.1364
OP2E10	8	43	95.45	0.0455	OP2E20	7	46	94.8	0.0519

Table A.19 Non-optimized trainings (with 3 modules, Voice, 80 %)

Training	Images for training (%)	Rec. Rate (%)	Error Rec.	Training	Images for training (%)	Rec. Rate (%)	Error Rec.
T1V1	67	88.74	0.1126	T1V16	11	77.06	0.2294
T1V2	17	80.68	0.1932	T1V17	10	60.17	0.3983
T1V3	51	87.79	0.1221	T1V18	73	91.77	0.0823
T1V4	6	71.57	0.2843	T1V19	34	90.72	0.0928
T1V5	39	77.27	0.2273	T1V20	36	88.31	0.1169
T1V6	48	86.23	0.1377	T1V21	28	84.42	0.1558
T1V7	13	49.78	0.5022	T1V22	19	71.75	0.2825
T1V8	39	88.31	0.1169	T1V23	45	82.60	0.1740
T1V9	41	89.39	0.1061	T1V24	58	94.81	0.0519
T1V10	53	88.83	0.1117	T1V25	53	93.25	0.0675
T1V11	36	88.53	0.1147	T1V26	55	93.83	0.0617
T1V12	23	79.06	0.2094	T1V27	38	84.42	0.1558
T1V13	66	95.67	0.0433	T1V28	50	95.84	0.0416
T1V14	8	69.55	0.3045	T1V29	23	76.95	0.2305
T1V15	20	57.79	0.4221	T1V30	57	91.88	0.0812

Table A.20 Non-optimized trainings (with different number of modules, Voice, 80 %)

Training	Modules	Images for training (%)	Rec. Rate (%)	Error Rec.	Training	Modules	Images for training (%)	Rec. Rate (%)	Error Rec.
T2V1	8	46	87.79	0.1221	T2V16	3	49	88.05	0.1195
T2V2	3	72	91.77	0.0823	T2V17	7	76	88.96	0.1104
T2V3	9	32	90.17	0.0983	T2V18	9	73	95.24	0.0476
T2V4	4	57	91.88	0.0812	T2V19	8	70	96.10	0.0390
T2V5	3	58	91.23	0.0877	T2V20	10	9	77.92	0.2208
T2V6	9	64	93.18	0.0682	T2V21	10	70	97.40	0.0260
T2V7	3	41	90.04	0.0996	T2V22	3	70	90.48	0.0952
T2V8	3	56	89.94	0.1006	T2V23	5	36	90.04	0.0996
T2V9	10	38	92.86	0.0714	T2V24	5	13	61.18	0.3882
T2V10	2	37	86.36	0.1364	T2V25	3	65	91.34	0.0866
T2V11	6	46	91.95	0.0805	T2V26	2	78	96.75	0.0325
T2V12	6	12	78.07	0.2193	T2V27	3	9	52.81	0.4719
T2V13	2	28	82.19	0.1781	T2V28	5	59	91.56	0.0844
T2V14	6	46	89.09	0.1091	T2V29	1	42	87.66	0.1234
T2V15	6	8	66.52	0.3348	T2V30	8	36	90.69	0.0931

Table A.21 Training non-optimized (with 3 modules, Voice, 50 %)

Training	Images for training (%)	Rec. Rate (%)	Error Rec.	Training	Images for training (%)	Rec. Rate (%)	Error Rec.
T3V1	42	83.98	0.1602	T3V16	7	77.06	0.2294
T3V2	11	73.30	0.2670	T3V17	7	60.17	0.3983
T3V3	33	85.16	0.1484	T3V18	46	91.43	0.0857
T3V4	5	71.57	0.2843	T3V19	22	82.31	0.1769
T3V5	25	83.12	0.1688	T3V20	23	82.47	0.1753
T3V6	30	84.60	0.1540	T3V21	18	72.89	0.2711
T3V7	9	49.78	0.5022	T3V22	12	62.34	0.3766
T3V8	25	82	0.1800	T3V23	29	81.08	0.1892
T3V9	26	84.97	0.1503	T3V24	37	95.24	0.0476
T3V10	34	86.09	0.1391	T3V25	34	90.91	0.0909
T3V11	23	85.39	0.1461	T3V26	35	90.48	0.0952
T3V12	15	79.06	0.2094	T3V27	24	78.25	0.2175
T3V13	42	86.80	0.1320	T3V28	32	88.13	0.1187
T3V14	6	69.55	0.3045	T3V29	15	76.95	0.2305
T3V15	13	38.24	0.6176	T3V30	36	84.42	0.1558

Table A.22 Non-optimized trainings (with different number of modules, Voice, 50 %)

Training	Modules	Images for training (%)	Rec. Rate (%)	Error Rec.	Training	Modules	Images for training (%)	Rec. Rate (%)	Error Rec.
T4V1	10	22	91.72	0.0828	T4V16	10	37	90.04	0.0996
T4V2	8	29	90.35	0.0965	T4V17	7	28	90.35	0.0965
T4V3	10	24	81.49	0.1851	T4V18	3	38	89.18	0.1082
T4V4	10	23	86.36	0.1364	T4V19	5	28	84.60	0.1540
T4V5	9	8	72.15	0.2785	T4V20	7	49	95.06	0.0494
T4V6	8	45	96.10	0.0390	T4V21	9	27	90.54	0.0946
T4V7	8	48	96.62	0.0338	T4V22	9	42	92.86	0.0714
T4V8	10	6	76.48	0.2352	T4V23	9	45	93.77	0.0623
T4V9	6	4	81.67	0.1833	T4V24	9	18	88.64	0.1136
T4V10	5	21	85.88	0.1412	T4V25	4	14	64.65	0.3535
T4V11	3	6	66.23	0.3377	T4V26	8	43	88.53	0.1147
T4V12	2	38	88.53	0.1147	T4V27	7	19	84.58	0.1542
T4V13	3	5	63.49	0.3651	T4V28	8	23	86.20	0.1380
T4V14	7	17	77.11	0.2289	T4V29	2	3	49.21	0.5079
T4V15	5	27	86.27	0.1373	T4V30	10	9	81.82	0.1818

Table A.23 Optimized results (Voice, 80 %)

Evolution	Modules	Images for training (%)	Rec. Rate (%)	Error Rec.	Evolution	Modules	Images for training (%)	Rec. Rate (%)	Error Rec.
OP1V1	7	72	97.84	0.0216	OP1V11	5	39	96.54	0.0346
OP1V2	9	55	98.70	0.0130	OP1V12	7	46	98.18	0.0182
OP1V3	4	66	99.13	0.0087	OP1V13	9	79	100	0
OP1V4	5	75	100	0	OP1V14	6	60	100	0
OP1V5	8	79	100	0	OP1V15	10	71	100	0.0000
OP1V6	10	55	99.68	0.0032	OP1V16	8	72	99.57	0.0043
OP1V7	9	73	100	0	OP1V17	5	65	99.13	0.0087
OP1V8	3	78	100	0	OP1V18	8	77	100	0
OP1V9	10	77	100	0	OP1V19	8	72	100	0
OP1V10	7	72	100	0	OP1V20	6	73	100	0

Table A.24 Optimized results (Voice, 50 %)

Evolution	Modules	Images for training (%)	Rec. Rate (%)	Error Rec.	Evolution	Modules	Images for training (%)	Rec. Rate (%)	Error Rec.
OP2V1	6	41	97.40	0.0260	OP2V11	4	37	96.32	0.0368
OP2V2	7	37	96.97	0.0303	OP2V12	9	35	96.97	0.0303
OP2V3	7	38	97.62	0.0238	OP2V13	8	17	93.67	0.0633
OP2V4	7	42	98.27	0.0173	OP2V14	7	36	97.40	0.0260
OP2V5	9	37	97.62	0.0238	OP2V15	6	35	97.62	0.0238
OP2V6	6	48	98.18	0.0182	OP2V16	9	49	98.96	0.0104
OP2V7	7	47	98.96	0.0104	OP2V17	8	49	98.18	0.0182
OP2V8	7	41	98.05	0.0195	OP2V18	7	48	98.96	0.0104
OP2V9	4	36	95.67	0.0433	OP2V19	3	50	98.18	0.0182
OP2V10	9	44	95.67	0.0433	OP2V20	7	35	97.84	0.0216

Table A.25 Non-optimized trainings (with different number of modules, Face (Essex), 80 %, color RGB)

Training	Modules	Images for training (%)	Rec. Rate (%)	Error Rec.	Training	Modules	Images for training (%)	Rec. Rate (%)	Error Rec.
T2C1F1	8	48	74.72	0.2528	T2C1F16	9	15	20.21	0.7979
T2C1F2	4	50	18.05	0.8195	T2C1F17	10	66	81.39	0.1861
T2C1F3	3	58	21.86	0.7814	T2C1F18	10	3	54.71	0.4529
T2C1F4	8	23	70.46	0.2954	T2C1F19	5	12	70.91	0.2909
T2C1F5	8	25	59.01	0.4099	T2C1F20	9	14	65.67	0.3433
T2C1F6	2	30	89.38	0.1062	T2C1F21	10	71	60.09	0.3991
T2C1F7	7	22	46.54	0.5346	T2C1F22	9	79	13.21	0.8679
T2C1F8	10	30	54.65	0.4535	T2C1F23	9	20	15.99	0.8401
T2C1F9	10	12	58.97	0.4103	T2C1F24	9	26	12.27	0.8773
T2C1F10	9	13	52.64	0.4736	T2C1F25	9	44	22.89	0.7711
T2C1F11	4	79	31.03	0.6897	T2C1F26	7	52	15.49	0.8451
T2C1F12	8	32	21.72	0.7828	T2C1F27	10	57	62.91	0.3709
T2C1F13	5	24	18.15	0.8185	T2C1F28	10	30	72.12	0.2788
T2C1F14	9	32	66.59	0.3341	T2C1F29	8	76	57.64	0.4236
T2C1F15	9	70	58.97	0.4103	T2C1F30	8	5	36.41	0.6359

Table A.26 Non-optimized trainings (with different number of modules, Face (Essex), 80 %, color RGB)

Training	Modules	Images for training (%)	Rec. Rate (%)	Error Rec.	Training	Modules	Images for training (%)	Rec. Rate (%)	Error Rec.
T2C2F1	9	74	92.00	0.0800	T2C2F16	9	55	55.50	0.4450
T2C2F2	8	31	28.68	0.7132	T2C2F17	9	13	11.95	0.8805
T2C2F3	9	37	92.86	0.0714	T2C2F18	6	54	69.29	0.3071
T2C2F4	6	63	95.46	0.0454	T2C2F19	6	22	33.65	0.6635
T2C2F5	7	61	79.42	0.2058	T2C2F20	8	16	47.06	0.5294
T2C2F6	8	42	94.66	0.0534	T2C2F21	5	67	81.39	0.1861
T2C2F7	5	47	25.59	0.7441	T2C2F22	6	28	55.82	0.4418
T2C2F8	2	37	17.40	0.8260	T2C2F23	5	72	65.90	0.3410
T2C2F9	10	50	90.10	0.0990	T2C2F24	9	3	17.65	0.8235
T2C2F10	10	37	83.63	0.1637	T2C2F25	2	68	32.65	0.6735
T2C2F11	3	74	55.59	0.4441	T2C2F26	5	4	22.89	0.7711
T2C2F12	8	21	51.96	0.4804	T2C2F27	8	36	68.09	0.3191
T2C2F13	3	64	50.33	0.4967	T2C2F28	2	48	29.54	0.7046
T2C2F14	4	75	58.05	0.4195	T2C2F29	5	20	39.87	0.6013
T2C2F15	2	76	34.77	0.6523	T2C2F30	6	2	14.66	0.8534

Table A.27 External Memory of Evolution #1

Granule	Validation					
	V_{s1}	V_{s2}	V_{s3}	V_{s4}	V_{s5}	Average
1	94.30 % 0.0570	96.96 % 0.0304	98.16 % 0.0184	97.94 % 0.0206	95.96 % 0.0404	96.66 % 0.0334
2	99.90 % 0.0010	100 % 0	99.87 % 0.0013	100 % 0	100 % 0	99.96 % 0.0004
3	100 % 0	100 % 0	100 % 0	100 % 0	100 % 0	100 % 0
Average	98.07 % 0.0193	98.99 % 0.0101	99.34 % 0.0066	99.31 % 0.0069	98.65 % 0.0135	98.87 % 0.0113

Table A.28 External Memory of Evolution #2

Granule	Validation					
	V_{s1}	V_{s2}	V_{s3}	V_{s4}	V_{s5}	Average
1	79.57 % 0.2043	95.76 % 0.0424	95.59 % 0.0441	97.55 % 0.0245	98.53 % 0.0147	93.40 % 0.0660
2	90.61 % 0.0939	99.55 % 0.0045	99.71 % 0.0029	100 % 0	100 % 0	97.97 % 0.0203
3	95.59 % 0.0441	100 % 0	99.90 % 0.0010	100 % 0	100 % 0	99.10 % 0.0090
Average	88.59 % 0.1141	98.44 % 0.0156	98.40 % 0.0160	99.18 % 0.0082	99.51 % 0.0049	96.82 % 0.0318

Table A.29 External Memory of Evolution #3

Granule	Validation					
	V_{s1}	V_{s2}	V_{s3}	V_{s4}	V_{s5}	Average
1	94.90 % 0.0510	97.65 % 0.0235	97.65 % 0.0235	98.16 % 0.0184	98.53 % 0.0147	97.38 % 0.0262
2	99.73 % 0.0027	100 % 0	100 % 0	100 % 0	100 % 0	99.95 % 0.0005
3	99.91 % 0.0009	100 % 0	100 % 0	99.80 % 0.0020	100 % 0	99.94 % 0.0006
Average	98.18 % 0.0182	99.22 % 0.0078	99.22 % 0.0078	99.32 % 0.0068	99.51 % 0.0049	99.09 % 0.0091

Table A.30 External Memory of Evolution #4

Granule	Validation					
	V_{s1}	V_{s2}	V_{s3}	V_{s4}	V_{s5}	Average
1	91.44 % 0.0856	98.42 % 0.0158	97.99 % 0.0201	98.90 % 0.0110	97.90 % 0.0210	96.93 % 0.0307
2	100 % 0	99.88 % 0.0012	99.72 % 0.0028	100 % 0	100 % 0	99.92 % 0.0008
3	100 % 0	100 % 0	99.85 % 0.0015	100 % 0	100 % 0	99.97 % 0.0003
Average	97.15 % 0.0285	99.43 % 0.0057	99.19 % 0.0081	99.63 % 0.0037	99.30 % 0.0070	98.94 % 0.0106

Table A.31 External Memory of Evolution #5

Granule	Validation					
	V_{s1}	V_{s2}	V_{s3}	V_{s4}	V_{s5}	Average
1	85.99 % 0.1401	83.04 % 0.1696	97.06 % 0.0294	97.79 % 0.0221	98.53 % 0.0147	92.48 % 0.0752
2	91.90 % 0.0810	99.10 % 0.0090	99.34 % 0.0066	100 % 0	100 % 0	98.07 % 0.0193
3	98.73 % 0.0127	99.91 % 0.0009	99.88 % 0.0012	100 % 0	100 % 0	99.70 % 0.0030
Average	92.21 % 0.0779	94.02 % 0.0598	98.76 % 0.0124	99.26 % 0.0074	99.51 % 0.0049	96.75 % 0.0325

Appendix B
Appendix of Fuzzy Integration

The results of each case using the optimization of 3 MFs are presented in Table B.1. The results obtained using the proposed HGA for the fuzzy inference optimization are presented in Table B.2.

Table B.1 Optimized Fuzzy integration results (3 MFs)

Evolution	C1		C2		C3		C4		C5		C6	
	FL	%	FL	%	FL	%	FL	%	FL	%	FL	%
EV1	1	87.55	2	93.38	2	88.31	2	94.26	2	100	2	78.04
EV2	1	87.12	2	93.90	2	86.36	2	95.67	2	100	1	64.70
EV3	1	87.45	2	92.99	2	86.36	2	95.24	2	100	2	79.81
EV4	1	87.34	1	89.09	2	88.53	2	93.83	2	100	2	78.63
EV5	1	87.45	2	93.77	2	87.88	2	95.89	2	100	2	77.80
EV6	2	88.74	1	89.09	2	73.92	2	93.83	1	100	2	78.16
EV7	2	86.26	2	86.26	2	88.31	2	94.48	1	100	2	77.33
EV8	1	87.34	2	93.77	2	86.69	2	94.05	1	100	2	76.86
EV9	2	92.53	1	88.96	1	76.08	1	85.28	1	100	2	77.57
EV10	1	87.55	1	88.83	2	88.42	2	93.29	1	100	2	76.74
EV11	1	88.64	1	89.09	2	88.20	2	93.72	2	100	2	77.92
EV12	1	87.12	2	93.90	2	75.00	1	79.76	2	100	2	76.74
EV13	1	87.66	2	91.56	2	87.77	2	94.05	2	100	2	78.16
EV14	1	87.45	2	92.47	2	86.36	2	93.83	2	100	2	77.21
EV15	1	87.12	1	88.96	2	88.96	2	95.45	1	100	2	77.57
EV16	2	88.64	2	93.51	2	87.77	2	94.81	1	100	2	78.98
EV17	1	87.45	1	89.09	2	74.03	2	94.26	2	100	2	76.86
EV18	2	88.53	1	89.09	2	75.32	2	93.83	2	100	2	77.10
EV19	1	87.34	2	93.38	2	86.36	2	93.72	1	100	2	77.21
EV20	1	87.34	1	88.83	2	87.77	1	84.20	2	100	2	76.98

© The Author(s) 2016
D. Sanchez and P. Melin, *Hierarchical Modular Granular Neural Networks with Fuzzy Aggregation*, SpringerBriefs in Computational Intelligence, DOI 10.1007/978-3-319-28862-8

Table B.2 Optimized Fuzzy integration results (Proposed HGA)

Evolution	C1		C2		C3		C4		C5		C6	
	FL	%	FL	%	FL	%	FL	%	FL	%	FL	%
EV1	2	95.13	2	98.18	2	88.42	2	95.78	2	100	2	80.40
EV2	2	94.59	2	96.88	2	90.15	2	96.86	2	100	2	79.46
EV3	1	94.81	2	97.66	2	88.74	1	95.35	1	100	2	78.51
EV4	2	93.94	2	97.14	2	88.53	2	96.32	1	100	2	80.76
EV5	1	93.51	2	98.18	2	90.15	2	96.43	1	100	2	80.40
EV6	2	93.51	2	97.14	2	87.34	2	95.02	2	100	1	80.40
EV7	1	94.70	1	98.31	2	89.83	2	95.45	2	100	2	78.75
EV8	2	94.70	2	98.44	2	88.42	2	95.89	2	100	2	80.64
EV9	2	94.70	1	98.70	2	89.39	2	96.65	1	100	2	78.63
EV10	2	95.13	2	97.79	2	89.07	2	96.21	1	100	2	80.17
EV11	1	95.67	2	98.05	1	86.69	2	96.00	1	100	2	78.75
EV12	2	92.32	1	96.88	2	87.88	2	96.00	2	100	2	81.35
EV13	1	94.37	2	96.62	2	88.42	2	96.86	1	100	2	82.88
EV14	2	92.86	2	98.96	2	90.15	2	95.45	2	100	2	80.40
EV15	1	94.48	2	96.36	2	88.96	2	96.32	2	100	1	80.52
EV16	1	94.70	1	96.36	1	88.42	2	95.89	2	100	1	80.64
EV17	1	94.48	2	98.31	2	89.50	2	95.24	1	100	2	81.35
EV18	2	93.72	1	97.40	2	89.07	2	94.70	1	100	2	79.57
EV19	1	93.61	1	97.66	2	89.07	2	96.21	2	100	2	78.39
EV20	2	93.18	2	98.31	2	90.69	2	93.94	1	100	1	79.46

Index

© The Author(s) 2016
D. Sanchez and P. Melin, *Hierarchical Modular Granular Neural Networks with Fuzzy Aggregation*, SpringerBriefs in Computational Intelligence, DOI 10.1007/978-3-319-28862-8